Early Modern Nautical Charts of the Adriatic Sea

Josip Faričić • Tome Marelić
Editors

Early Modern Nautical Charts of the Adriatic Sea

Information Sources, Navigation Tools, and Communication Media

palgrave
macmillan

Editors
Josip Faričić
Department of Geography
University of Zadar
Zadar, Croatia

Tome Marelić
Department of Geography
University of Zadar
Zadar, Croatia

Croatian Science Foundation. This work was supported by the Croatian Science Foundation (IP-2020-02-5339 Early Modern Nautical Charts of the Adriatic Sea).

ISBN 978-981-96-8045-0 ISBN 978-981-96-8046-7 (eBook)
https://doi.org/10.1007/978-981-96-8046-7

This Palgrave Macmillan imprint is published by the registered company Springer Nature Singapore Pte Ltd.
The registered company address is: 152 Beach Road, #21-01/04 Gateway East, Singapore 189721, Singapore

Preface

As the outcome of the scientific research project *Early Modern Nautical Charts of the Adriatic Sea: Information Sources, Navigation Means, and Communication Media*, the book provides an extensive understanding of the history of nautical cartography for the Adriatic Sea and its coastal regions, from the appearance of the earliest known examples in the late thirteenth century to the introduction of systematic geodetic and hydrographic surveys in the late eighteenth and early nineteenth centuries.

The volume opens with a concise historical overview of the development of nautical cartography in the Adriatic Sea within the broader context of Mediterranean maritime traditions during the period under study. This introductory chapter also provides a critical synthesis of the most significant scientific contributions to the field, thereby establishing an editorial framework for the chapters that follow. These subsequent chapters examine a range of historiographical and phenomenological dimensions of nautical cartography, with particular attention to methodological approaches for analysing old nautical charts. Each chapter is conceived as a self-contained unit, enabling the book to be read both sequentially and selectively, according to the reader's interest.

By consulting a large number of primary sources, Patrizia Licini, Orietta Selva, Dragan Umek, Ahmet Üstüner, and Okan Büyüktapu provide comprehensive reviews of the historical context of the nautical cartography of the Venetian Republic and the Ottoman Empire—entities that at the time had the greatest impact on the political and cultural landscape of the Adriatic. Julijan Sutlović delves into the ways in which the numerous layers of information contained in late mediaeval and early modern nautical charts had an impact on their ability to communicate with their users. Lena Mirošević, Dubravka Mlinarić, and Josip Faričić further enhance this topic by conducting thorough analyses of the diverse range of symbolism encapsulated in these charts, encompassing both artistic and religious elements with complex semiotic implications. Milorad Pavić focuses on the isloarios, a publication genre that fused the short historiographic overviews and descriptions of coastal and island areas with charts and maps, while Tome Marelić emphasizes the key methodological elements of cartometric approaches for studying the geometry of old charts and maps, as well as the significance of incorporating the results of such analyses with the results that come from descriptive approaches to the subject matter.

Since the Adriatic Sea has always functioned as "a Mediterranean within a Mediterranean" throughout history, and because early modern nautical cartography absorbed many non-navigational elements and became intertwined with other branches of cartography, the concepts and methods presented in this book transcend both the geographical boundaries of the Adriatic and nautical cartography in itself in many ways. As a result, we hope that it will attract the attention of a variety of readers who have an interest in the history of cartography in a broader sense.

Josip Faričić
Tome Marelić

Acknowledgements

This book is based on the findings of the scientific research project IP-2020-02-5339 *Early Modern Nautical Charts of the Adriatic Sea: Information Sources, Navigation Means, and Communication Media* (NACHAS) funded by the Croatian Science Foundation.

HRZZ
Croatian Science
Foundation

NACHAS

Contents

1 **Early Modern Nautical Cartography of the Adriatic Sea: An Overview** 1
Josip Faričić and Tome Marelić

2 **Cultural Directions for the Adriatic Sea, or the Gulf of Venice** 37
Patrizia Licini de Romagnoli

3 **Early Modern Nautical Charts of the Adriatic Sea as a Medium of Communication** 67
Julijan Sutlović

4 ***Navigare et Regere*: Venetian Cartography Through the Centuries—From Planning Tool to Symbol of Power** 99
Orietta Selva and Dragan Umek

5 **Symbolism of Artistic Elements on Nautical Charts of the Adriatic Sea** 123
Lena Mirošević and Dubravka Mlinarić

6 **Religious Symbolism on Old Nautical Charts of the Adriatic Sea** 147
Josip Faričić

7 **Credibility of the Maps of the Adriatic Isolarii in the Sixteenth Century: Creation of Different Geographical Identities** 169
Milorad Pavić

8 **Mapping the Adriatic: Ottoman Nautical Charts Before the Treaty of Karlowitz** 193
Ahmet Üstüner and Okan Büyüktapu

9 **Cartometric Analyses: Methodological Issues and Their Significance for the History of Cartography and Historical Studies in General** 223
Tome Marelić

Appendix 249

Index 257

Notes on Contributors

Okan Büyüktapu is an assistant professor in the Department of History, Ege University, specializing in Early Modern Ottoman history and Ottoman--Croatian relations. He completed his PhD at the University of Zagreb (2022), focusing on Ottoman military and administrative history in Slavonia. His research interests encompass Ottoman military history, social and economic history, travel literature, cartography, and maritime history. Büyüktapu has published on Ottoman military organization in Croatia and Bosnia and Herzegovina, with emphasis on the Ottoman military frontier (serhad). His knowledge of both Croatian and Ottoman sources enables him to provide new insights into Croatia's Ottoman period.

Patrizia Licini de Romagnoli studied Modern History and Foreign Languages at Catholic University, Milan. She taught first at Bergamo University as Professor of Human Geography, before joining Macerata-Fermo University in 2000 as Professor of Geography of the Ancient World. She is currently a member of the Association of University Professors of Geography in Italy, and of the Mediävistenverband (e.V.). Licini has published widely on comparative history of cartography, and especially on East European countries and how they have related to the global world structure after the advent of the portolan chart (probably a little before 1300).

Josip Faričić is a professor in the Department of Geography at the University of Zadar, Croatia. His main areas of interest are cartography, historical geography, and the geography of islands. He analyses old maps as sources of spatial data which, together with archival and other sources, enable research into historical changes in spatial functions and thus the landscape. He is the author or co-author of more than a hundred papers published in scientific journals, proceedings, and books on cartography, historical and maritime geography, and geography of religions. From 2023 he has been the rector of the University of Zadar.

Tome Marelić is an assistant professor at the University of Zadar, Department of Geography. His scientific work focuses on the history of cartography, with his primary research interest being the still-mysterious origins of portolan charts, which he seeks to resolve by employing the cartometric approach in conjunction with known historical sources. He is continuously dedicated to scientific and research work, the results of which he regularly presents at international conferences and publishes in established scientific journals. Currently, he is the chair of the Croatian Cartographic Society's body on cartographic heritage.

Lena Mirošević is a professor in the Department of Geography at the University of Zadar. Her research focuses on historical cartography and geography as well as cultural geography. She has actively participated in numerous scientific conferences and academic events in Croatia and internationally. As both an independent author and co-author, she has made significant contributions to the field, with her research published in journals indexed in WoS and Scopus.

Dubravka Mlinarić is a research advisor at the Institute for Migration Research. She holds degrees in History and Geography from the Faculty of Humanities and Social Sciences and the Faculty of Science at the University of Zagreb. She earned her master's degree in History from the Central European University in Budapest, and a PhD in History from the University of Zagreb and has published more than sixty publications. Since 2008 she has been teaching the course Cartographic Sources for Premodern History. Her research interests include early modern history within a regional and interdisciplinary context, emphasizing historical demography and migrations, cartography, and environmental history.

Milorad Pavić is a full professor at the University of Zadar, Department of History. Since 1998, he has participated in multiple scientific projects. His scientific interests cover a wide range of topics of world and national history, but his primary focus lies in the modern history of the eastern Adriatic, especially geographical, cartographic, and navigation manuals pertaining to the eastern Adriatic created by the late eighteenth century.

He has presented the results of his scientific research at numerous international conferences and in scientific journals. He is the author of two university textbooks and co-editor of a monograph.

Orietta Selva is an Associate Professor of Geography at the University of Trieste, Department of Humanities (DiSU). Her cultural activities and academic research deal with cartography and historical geography, teacher education, and human geography. During her career, she has organized and set up various exhibitions of ancient cartography regarding the area of the Northern Adriatic Sea. She has been working both in Italy and abroad together with collectors, antique dealers, institutions, organizations, and associations towards the collection, preservation, and dissemination of cartographic heritage.

Julijan Sutlović is a Postdoctoral Researcher at the University of Zadar, Department of Geography. At the moment his research focuses on cartography and the history of cartography, with a particular interest in early modern nautical charts. He has published in internationally recognized scientific journals and actively presents his research at international conferences.

Dragan Umek is an associate professor in the Department of Humanities (DiSU) of the University of Trieste, where he teaches Geography. His interest centres on historical cartography, migrations, and human and cultural geography with particular regard to the Adriatic area and the Balkan Region. He is engaged in the constant implementation and valorization of the ancient cartography heritage of the department within the Centre for Geographical and Territorial Research. As co-author, he has set up numerous ancient map exhibitions, as well as edited catalogues and books on geo-cartographical topics.

Ahmet Üstüner is an associate professor at Istanbul University's Research Institute of Turkology. He completed his PhD at the same university with a dissertation on the famous Ottoman geographer Katib Çelebi's Ottoman translation of the Mercator Atlas. His research interests include the history of the Ottoman Empire in the early modern and classical ages, with a particular focus on Ottoman and European cartography and Ottoman geographical works. He is currently working on a book about the Ottoman translation of the Blaeu Atlas. He also carries out projects related to publishing Turkish-Islamic geographical works and Ottoman medical history texts.

List of Figures

Fig. 1.1	Political map of South and Southeast Europe for the year 1600. Basemap shapefile source: marineregions.org (Claus et al. 2017)	2
Fig. 1.2	Political map of South and Southeast Europe for the year 1750. Basemap shapefile source: marineregions.org (Claus et al. 2017)	3
Fig. 1.3	Milestones in the history of nautical cartography for the Adriatic Sea area	5
Fig. 1.4	Pietro Vesconte, nautical chart of the Adriatic Sea, Venice, 1318—the first manuscript nautical chart of the Adriatic Sea (Source: Austrian National Library, Vienna, Call Number: Cod. 594 (Cimel. 20), 10v-11r)	9
Fig. 1.5	Bartolomeo Zamberti da li Soneti, nautical chart of the Adriatic Sea, Venice, 1485—an example of a chart that is part of comprehensive publications (Source: National Maritime Museum, Greenwich, London; P/21(2); MS 38-9920C)	10
Fig. 1.6	The nautical chart of the northern part of the Central Mediterranean from Alonso de Santa Cruz's *Islario general de todas las islas del mundo* (1539–1560) containing the markings of latitudes and longitudes (Source: National Library of Spain (Biblioteca Nacional de España), Madrid, MSS. MICRO/12638)	11
Fig. 1.7	Willem Barentsz, nautical chart of the Adriatic Sea, Amsterdam, 1595—prototype of printed nautical charts of the Adriatic Sea (Source: Stanford University Libraries, The Barry Lawrence Ruderman Map Collection, Stanford)	12

Fig. 1.8 Muhiddin Pîrî Reis, nautical chart of the Adriatic Sea, Gallipoli, 1526 (Source: The Walters Art Museum, Baltimore; W.658, fol. 208a) 13

Fig. 1.9 Four portolan charts of the Adriatic Sea with nearly identical coastline renderings: (**a**) Battista Agnese, 1538 (Source: University of Pennsylvania, Kislak Center for Special Collections, Rare Books and Manuscripts, Philadelphia; LJS 28), (**b**) Diogo Homem, 1570 (Source: Croatian State Archives, Cartographic Collection, Zagreb; HR-HDA-902, D. XIV.6), (**c**) Vincus Demetrius Voltius, 1593 (Source: National Library of Finland, Maps, The Nordenskiöld Map Collection, Helsinki; N-Kt-103c), (**d**) Marcheto Fassoi, 1679 (Source: The Huntington Library, Art Museum, and Botanical Gardens, Library Collections, Maps and Atlases, Portolans, San Marino, CA, USA; mssHM 30) 15

Fig. 1.10 Jacques-Nicolas Bellin, *Carte hydrographique du Golphe de Venise*, Paris, 1771, containing magnetic declination data for the Adriatic Sea area; see enlarged circular inset (Source: National and University Library, Map and Atlas Collection, Zagreb; S-JZ-XVIII-145) 16

Fig. 1.11 Vincenzo Maria Coronelli's nautical chart (1690–1694) of the Sazan island (Chart source: Vincenzo Maria Coronelli, Isolario, Vol. IV, Venice, 1690–1694 (Source: State Archives in Zadar, Library, Zadar; II.A*) 17

Fig. 1.12 Lodovico Furlanetto, *Nuova carta Marittima del Golfo di Venezia*, Venice, 1784—a medieval nautical cartographic image of the Adriatic Sea in a printed edition from the end of the early modern period (Source: State Archives in Zadar, Cartographic collection, Zadar; HR-DAZD-383 No. 3.1) 17

Fig. 2.1 Ordinance of the Cesarean Royal Major-General in Pula and commander of the Vodnjan garrison, Count von Klenau, dated June 19, 1797, for Vodnjan (Source: Trieste, Archivio di Stato, Atti amministrativi dell'Istria 1797–1813, busta 47, c. 650–651). Courtesy of the Library (09.09.2015) 40

Fig. 2.2 Bartolomeo Pareto's 1455 luxury nautical chart (Source: Rome, Biblioteca Nazionale Centrale, Carte nautiche 1) © BNC Board 43

Fig. 2.3 Map of the region around the Adriatic, from an atlas by Angelo Freducci (Source: Mantua, Biblioteca Comunale Teresiana, MS 646, Map 6; atlas dated 1556). Courtesy of the Library (18/2019) 47

Fig. 2.4 Bonomini's 1570 chart (Source: Roma, Galleria Colonna, Tabula Nautica). Courtesy of the Galleria Colonna, Rome 50
Fig. 2.5 Corte named Licini o del piombo (lead ore), Venice, Sestiere Castello, Rio de San Zulian, Calle de la Fava. Detail of the temporary storage rooms. The *Poglizzani* assume monopoly for the transportation of lead ore from Šibenik to Venice (Venice and Zadar, 1765, Provveditore Generale: Pietro Michiel). The family is called Licini in Zadar and Venice, and Rubcich in *Poglizza* 56
Fig. 3.1 Koláčnýs's communication of cartographic information model (Source: Koláčný 1969: 41) 72
Fig. 3.2 Antonio Millo, nautical chart of the Adriatic Sea, Venice, 1583 (Source: Biblioteka Narodnowa, Warsaw, BN ZZK 0.2399) 76
Fig. 3.3 The personification of Venice; Vincenzo de Lucio, ca. 1792–1796. (Source: Private collection Marco Asta, Bologna) 77
Fig. 3.4 The hierarchy of toponyms: coloured in red on a manuscript chart: (**a**) Diogo Homem, 1570, and use of capitalization on a printed chart; (**b**) Lodovico Furlanetto, 1784 (Sources: (a) Hrvatski državni arhiv, Kartografska zbirka, Zagreb, HR-HDA-902, D.XIV.6; (b) Državni arhiv u Zadru, Kartografska zbirka, Zadar, HR-DAZD-383 No. 3.1) 79
Fig. 3.5 Depictions of Lokrum island on early modern nautical charts of the Adriatic: (**a**) Lokrum named as *I. Cromma*; Pieter Goos, 1650; (**b**) Lokrum named as I. *St. Marco*; Joseph Roux, 1764 (Sources: (a) Private Collection Marco Asta, Bologna; (b) Nacionalna i sveučilišna knjižnica u Zagrebu, Zbirka zemljovida i atlasa, Zagreb, S-JZ-XVIII-116) 81
Fig. 3.6 The exemplary sample of cartographic signs on nautical charts of the Adriatic Sea. Sources: Unknown author, ca. 1270, (Bibliothèque nationale de France, Paris, GE B-1118 (RES)); Pietro Vesconte, 1318, (Österreichische Nationalbibliothek, Vienna, Cod. 594 (Cimel. 20), 10v-11r); 9. Zuan Soligo, ca. 1489 (British Library, London, Egerton MS 73); Pietro Coppo, 1525 (Pomorski muzej Sergej Mašera, Piran, No. 1002307, T. 15); Antonio Millo, 1583 (Biblioteka Narodnowa, Warsaw, BN ZZK 0.2399); Willem Barents, 1595 (Stanford University Libraries, The Barry Lawrence Ruderman Map Collection, Stanford, Tooley, A-D:85, 268; K-P:10); Joseph Roux, 1764 (Nacionalna i sveučilišna knjižnica u Zagrebu, Zbirka zemljovida i atlasa, Zagreb, S-JZ-XVIII-116); Vincenzo de Lucio, 1809 (Sveučilište u Zadru, Znanstvena knjižnica, Zadar, 15188 D-20) 84

Fig. 3.7 Scales and units of distance on Barents's nautical chart of the Adriatic Sea, 1595 (Source: Stanford University Libraries, The Barry Lawrence Ruderman Map Collection, Stanford, Tooley, A-D:85, 268; K-P:10) 87

Fig. 3.8 Snippets from charts on which Premuda, Škarda, Ist, and Molat are represented: (**a**) P. Vesconte, 1318; (**b**) G. Giroldi, 1426; (**c**) V. de Lucio ca. 1792–1796; (**d**) Nautical chart Silba—Pag 100–19, 2003 (Sources: (a) Österreichische Nationalbibliothek, Vienna, Cod. 594 (Cimel. 20), 10v-11r; (b) Biblioteca Nazionale Marciana, Venice, It. VI, 212 (=5694); (c) Hrvatski državni arhiv, Kartografska zbirka, Zagreb, T 3 B + 5 D—HR-HDA-902 E.IV.13-a; (d) marineregions.org, obtained on April 23, 2025) 90

Fig. 4.1 Gaspare Tentivo, *Carta Marittima del GOLFO DI VENEZIA con tutte l'Isole e Scogli che in esso si trovano*, Venice, printed in c. 1740 (Source: Private collection Marco Asta, Bologna) 105

Fig. 4.2 Paolo Forlani, *IL GOLFO DI VENETIA*, Venice, 1568 (Source: Department of Humanities, University of Trieste) 107

Fig. 4.3 Abbate Lonati, *GOLFO DI VENEZIA*, eighteenth century (Source: Private collection Italico Stener) 107

Fig. 4.4 Stefano Scolari, *GOLFO DI VENETIA OVERO MARE ADRIATICO*, Venice, the end of the sixteenth century (Source: Private collection Italico Stener) 110

Fig. 4.5 Cristoforo Sorte, *Dissegno dilla Patria dil Friuli*…, Venice, 1590 (Source: Staatsarchiv-Kriegsarchiv, Vienna, KA/KS, B VII a 167) 114

Fig. 4.6 Francesco Leschiutta, *TORENTE LVMIEI*, 1769 (Source: ASVe, Provveditori Sopra Boschi, b. 170, dis. 11) 115

Fig. 4.7 Giacomo Spinelli, *DISEGNO DEL POSTO DELLA CHIUSA*, 1703 (Source: ASVe, Raccolta Terkuz, dis. 45 A) 117

Fig. 5.1 Coronelli's map of the Adriatic Sea, Venice, 1688 (Source: National and University Library, Map and Atlas Collection, Zagreb, Call number S-JZ-XVII-56) 129

Fig. 5.2 Unknown author, Depiction of the Adriatic, Dieppe, 1547 (Source: Berkeley Digital Library, Henry E. Huntington Library and Art Gallery, San Marino, CA, HM 29, f. 14) 132

Fig. 5.3 Unknown author, nautical chart of the Adriatic Sea, in: Dauphin atlas, Dieppe, c. 1538–1546 (Source: Koninklijke Bibliotheek, Nationale bibliotheek, The Hague; 129 A 24, 29v-30r) 134

Fig. 5.4 Depiction of the sea monsters on the nautical charts of the Adriatic Sea: (**a**) segment of Willem Barents' nautical chart,

159 (Source: The Barry Lawrence Ruderman Map Collection, Stanford University Libraries, Tooley, A-D:85, 268; K-P:10); (**b**) segment of Joseph Roux's nautical chart, Marseille, 1764 (Source: National and University Library, Map and Atlas Collection, Zagreb, S-JZ-XVIII-116) 137
Fig. 5.5 The depiction of the Adriatic Sea on Joan Martines' nautical chart of the eastern Mediterranean, c. 1578 (Source: The Huntington Library, Art Museum, and Botanical Gardens, Library Collections, Maps and Atlases, Portolans, San Marino, CA, USA, mssHM 33) 139
Fig. 5.6 Niccolò Guidalotti, nautical chart of the Adriatic Sea, in: Nautical Atlas, Venice 1646 (Source: Biblioteca Nazionale Marciana, It. IV, 10=5062; f. 2) 140
Fig. 6.1 Sebastiano Condina, nautical chart of the Mediterranean Sea, Naples, 1615 (Source: *Biblioteca Nazionale Marciana*, It. IV, 505 (=10036), t. 1) 150
Fig.6. 2 Compass roses with the fleur-de-lis marking north and the sign of the cross marking east on a nautical chart of the Adriatic Sea: (**a**) chart made by Diogo Homem (Venice, 1559) (Source: *Bibliothèque nationale de France, département Cartes et plans*, Paris; CPL GE DD-2003 (RES)); (**b**) chart made by Vincenzo Demetrio Volčić (Naples, 1593.) (Source: *National Library of Finland*, Maps, The Nordenskiöld Map Collection, Helsinki; N-Kt-103c) 153
Fig. 6.3 Compass rose on which east is indicated by a crescent moon on the seventh sheet of the *Carta di cabotaggio del Mare Adriatico, 1822–1824* (Source: State Archives in Zadar, Cartographic collection, Zadar; HR-DAZD-383 No. 3.2) 156
Fig. 6.4 Details featuring the depiction of the Winged Lion of St Mark: (**a**) Alvise Gramolin, Carta nautica del mar Adriatico, Venice, 1624 (Source: Museo Correr, Gabinetto di Cartografia, Cl. XLIVa, n. 0044); (**b**) Jean François Roussin, Carta du Golfo di Venetia, Venice, 1661 (Source: The Huntington Library, Art Museum, and Botanical Gardens, Library Collections, Maps and Atlases, Portolans, San Marino, CA, USA; mssHM 37) 158
Fig. 6.5 Unknown Author, nautical chart of the Central Mediterranean, late fifteenth century (Source: *Biblioteca Nazionale Marciana*; It. IV, 9 (=5090), t. 2) 158
Fig. 6.6 Composites of Christian and Ancient Greek and Roman religious motifs: (**a**) Mattheus Greuter, Sheet 2 of the Map of Italy, Detail, Bologna, 1630 (Source: *Bibliothèque nationale*

de France; GE DD-5867 (1-12RES), (**b**) Vincenzo Maria Coronelli, Map of the Dalmatia, Detail, Venice, ca 1690 (Source: Maritime and History Museum of the Croatian Littoral, Rijeka; PPMHP 122117) 159
Fig. 6.7 Pierre Collin, nautical chart of the Adriatic Sea, Marseille, 1642 (Source: *Bibliothèque Municipale de Lyon*, Lyon; MS 177) 161
Fig. 6.8 Pierre Van der Aa, Detail from the chart Golfe de Venise avec les Côtes maritimes, Bayes et Ports etc. De la Gréce, Dalmatie et Italie, Leiden, 1720 (Source: *Croatian State Archives, Cartographic Collection*, Zagreb; HR-HDA-902, E.IV.29) 162
Fig. 6.9 Ioannes van Keulen, detail from the chart Pas-Caart van de Weder zytsche Zee-kusten soo van Italia als Dalmatia en Griecken Inde Golff van Venetien, Amsterdam, 1700 (Source: *Croatian State Archives, Cartographic Collection, Zagreb*; HR-HDA-902, E.IV.30) 163
Fig. 7.1 *Parte de Schiavonia* map from Benedetto Bordone's *Isolario* (1547) showing the island and mainland parts of the Kvarner Gulf and North Dalmatia (Source: University of Zadar, Research Library, R 317) 173
Fig. 7.2 The map *ISOLA DI PAGO* published in Giuseppe Rosaccio's isolario *Viaggio da Venetia a Costantinopoli Per Mare, e per Terra, insieme quello di Terra Santa*, Venice, 1598 (Source: National and University Library, Zagreb, Collection of Manuscripts and Old Books, Zagreb; R-VI-16°-32) 175
Fig. 7.3 The distinction between the two quadrilateral polygons possessing identical areas, yet differing in the length-to-width ratio of their sides (a:b)—the polygon on the right exhibits a longer perimeter, potentially giving the impression of being larger when measured by the time required to circumnavigate it (if it were an island) 177
Fig. 7.4 North Dalmatia on the map *Zarra et Contado citta principale della Dalmatia posta sul mare adriatico locho delli Ill[ustrissi]mi S[igno]ri Venetiani al p[rese]nte molestada da Turchi* from Giovani Francesco Camocio's isolario, Venice, 1571 (Source: Archaeological Museum in Split; 45-e-39) 178
Fig. 7.5 The comparison of the Church of St George on Cape Marjan (encircled in red) on two different maps from Giuseppe Rosaccio's *Viaggio da Venetia, a Costantinopoli Per Mare … isolario* (1606): (**a**) segment of the map of Trogir and its surroundings on the left; and (**b**) segment of the city of Split with its surroundings on the right (Source: National and

University Library, Collection of Manuscripts and Old Books, Zagreb; R-VI-16°-32) 180
Fig. 7.6 The island of Korčula in Giuseppe Rosaccio's isolario, 1606 (Source: National and University Library, Collection of Manuscripts and Old Books, Zagreb; R-VI-16°-32) 182
Fig. 7.7 Giovanni Francesco Camocio's map of the island of Korčula (1571), most likely used by Giuseppe Rosaccio as a template (Source: National and University Library, Collection of Maps and Atlases, Zagreb; SJZXVI-40) 183
Fig. 8.1 Ahmed the Tangier, Mediterranean portolan chart from 1413 (Source: Topkapı Palace Museum Library, Hazine Collection, Manuscript No. 1823) 200
Fig. 8.2 Details of the Adriatic Section of Ahmed the Tangier's portolan chart from 1413 (Source: Topkapı Palace Museum Library, Hazine Collection, Manuscript No. 1823) 201
Fig. 8.3 Details of the Adriatic Section of Ibrahim al-Mursi's Nautical Chart from 1461 (Source: Istanbul Naval Museum, Manuscript No. 882) 202
Fig. 8.4 Ptolemy's *Kitâbü'l-Coğrâfya fi'l-Ma'mûre mine'l-Arz* (Source: Presidency of Manuscript Institution of Turkey, Süleymaniye Manuscript Library, Ayasofya Collection, Manuscript No. 2610, folio 19) 203
Fig. 8.5 Details of Venice (**a**) and the middle and southern parts of the Dalmatian coast and islands (**b**) from Piri Reis's *Kitâb-ı Bahriye* (1521) (Source: Presidency of Manuscript Institution of Turkey, Süleymaniye Manuscript Library, Ali Emiri Manuscript Library, Coğrafya Collection, Manuscript No. 1, folia 78, 80) 206
Fig. 8.6 Piri Reis, Venice, in *Kitâb-ı Bahriye* (1521) (Source: Presidency of Manuscript Institution of Turkey, Süleymaniye Manuscript Library, li Emiri Manuscript Library, Coğrafya Collection, Manuscript No. 1, folio 81) 207
Fig. 8.7 The Adriatic and Tyrrhenian Seas on two portolan charts; (**a**) from the *Walters Nautical Atlas* made in the 1560s (Source: Walters Collection, Manuscript No. W660, folio 7); (**b**) Ali Macar Reis's chart made in the sixteenth century (Source: Topkapı Palace Museum Library, Hazine Collection, Ali Macar Reis Atlası, Manuscript No. 644, folio 4) 209
Fig. 8.8 "Adriatic and Italy," *The Muhit Atlas* (Source: Biblioteca Nazionale di Napoli, BNN Ms. III.F.35, folio 99) 211
Fig. 8.9 Depiction of the central and southern parts of the Dalmatian coast and islands along with Dubrovik and Korčula in

Müntehab-ı Bahriye (Source: Presidency of Manuscript Institution of Turkey, Süleymaniye Manuscript Library, Aşir Efendi Collection No. 227, folio 62b) 212
Fig. 8.10 The depiction of the Adriatic Sea (**a**) and the depiction of North Dalmatia and the Croatian and Bosnian hinterland (**b**) in *Levâmi'u'n-Nûr fî Zulmet-i Atlas Minor* (Source: Presidency of Manuscript Institution of Turkey, Süleymaniye Manuscript Library, Nuruosmaniye Collection, No. 2998 folia 318a, 359a) 215
Fig. 8.11 Ebûbekir Behrâm b. ed-Dımeşkî, "Sûret-i İtalya," *Nusretü'l-İslam ve's-Sürûr fî Tahrîr-i Atlas Mayor* (Source: Istanbul University Central Library Rare Works, Turkish Manuscripts, 6606, folia 212b-213a) 217
Fig. 9.1 The simplified schematic of the georeferencing process and the accuracy estimation of old maps and charts by cartometric means (Portolan chart source: *Bibliothèque nationale de France*, GE B-1118 (RES)) 228
Fig. 9.2 The magnitude and orientation of displacement vectors of residuals of the anonymous *Carte Pisane* (c. 1270, the oldest known portolan chart) georeferenced to a modern map in the Mercator projection across the Mediterranean area. Basemap shapefile source: marineregions.org (Claus et al. 2017) 229
Fig. 9.3 The manually assembled composite of two Pietro Vesconte's portolan charts georeferenced to the modern map in the close approximation of the *Ptolemy I* projection centred near Gibraltar (the upper part), and to the modern map in the Mercator projection (the lower part), with their vectorised coastlines and de-projected accuracy values. The greater of the two axial values in bold indicate the estimated planimetric accuracy of the entire georeferenced unit (Portolan chart sources: *Archivio di Stato di Firenze*, CN 01; *Bibliothèque nationale de France*, CPL GE DD-687 (RES)). Basemap shapefile source: marineregions.org (Claus et al. 2017) 230
Fig. 9.4 The displacement vectors of residuals of the geometric centres of coordinates on three geographical maps (georeferenced by using the coordinates extracted from them) in comparison to Claudius Ptolemy's coordinates of locations on the Adriatic Sea coast. All datasets were longitudinally shifted by $\Delta\lambda_0 = -25.2°$. Ptolemy's coordinates source: Stevenson 1991. Basemap shapefile source: marineregions.org (Claus et al. 2017) 237

Fig. 9.5 Vincenzo Maria Coronelli's map from 1688 georeferenced to a modern map in the equidistant cylindrical projection $\varphi_0 = 36°$ (the upper part), and the composite of Robert Dudley's three nautical charts from his 1646 book *Dell'arcano del mare* georeferenced to a modern map in the Mercator projection $\varphi_0 = 36°$ (the lower part). Both units were georeferenced by using the coordinates extracted from them and with the application of proprietary longitudinal shift ($\Delta\lambda_0$) (Old map and nautical chart sources: *Nacionalna sveucilišna knjižnica u Zagrebu, zbirka karata i atlasa*, S-JZ-XVII-56; *National Library of Finland, Nordenskiöld collection* NBN:fi-fe201002051338). Basemap shapefile source: marineregions.org (Claus et al. 2017) 239
Fig. 9.6 Individually georeferenced small- and large-scale sheets from Battista Agnese's portolan atlas (1538), overlay of their vectorised coastlines with the coastlines of their coherent composite, and the magnetic declination for the year 1500 according to the CALS3k.4 paleomagnetic model (Portolan atlas source: *University of Pennsylvania, Rare Book & Manuscript Library*, LJS 28). Basemap shapefile source: marineregions.org (Claus et al. 2017). CALS3k.4 data source: GEOMAGIA50.v3.2, accessed: March 18, 2023 241
Fig. 9.7 Angelino Dulcert's portolan chart from 1339 georeferenced across the Mediterranean and Black Sea (black line) and its eight cartometrically determined subsections (colour-coded lines) with an average planimetric accuracy twice as high (Portolan chart source: *Bibliothèque nationale de France*, GE B-696 (RES)). Basemap shapefile source: marineregions.org (Claus et al. 2017) 242

List of Tables

Table 5.1 Maps and nautical charts depicting the Adriatic Sea used in the research 126
Table 7.1 Toponymic doublets in various isolarii texts and maps 185

CHAPTER 1

Early Modern Nautical Cartography of the Adriatic Sea: An Overview

Josip Faričić and *Tome Marelić*

1 Nautical Cartography of the Adriatic Sea in the Early Modern Period

Due to its geographic position and importance within the Mediterranean socioeconomic framework, the Adriatic Sea and its adjacent regions have historically been a locus of interactions and conflicts, as well as a hub for continuous cross-cultural exchanges of goods, ideas, and technologies, rendering it the "the Mediterranean on a smaller scale" (Braudel 1972; Ivetic 2019). In the early modern period, numerous political entities that significantly influenced Europe's political, economic, and cultural landscape also impacted the historic development of the Adriatic Sea (Figs. 1.1 and 1.2). The Republic of Venice, established in the Middle Ages and dissolved by Napoleon in 1797, was one of the most influential entities, consistently endeavouring to assert itself as the dominant power while adeptly navigating various political fluctuations. Alongside its considerable territorial expansion along the northern and eastern Adriatic coast, Venice has

J. Faričić (✉) • T. Marelić
Department of Geography, University of Zadar, Zadar, Croatia
e-mail: jfaricic@unizd.hr; tmarelic@unizd.hr

J. Faričić, T. Marelić (eds.), *Early Modern Nautical Charts of the Adriatic Sea*, https://doi.org/10.1007/978-981-96-8046-7_1

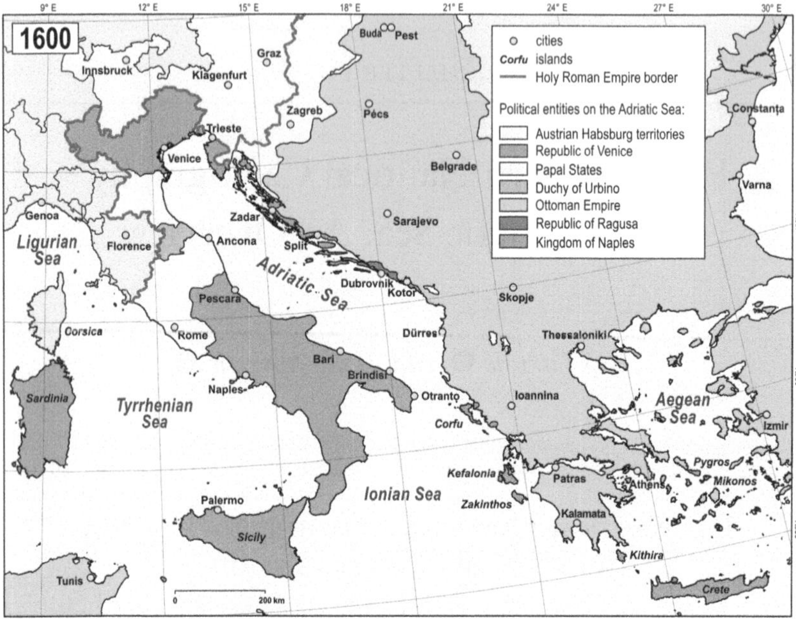

Fig. 1.1 Political map of South and Southeast Europe for the year 1600. Basemap shapefile source: marineregions.org (Claus et al. 2017)

governed Corfu since 1204, located at the junction of the Adriatic and Ionian Seas. Venetian official cosmographer Vincenzo Maria Coronelli stated in his isolario *Citta, Fortezze, ed altri Luoghi principali dell' Albania, Epiro e Livadia, E particolarmente i posseduti da Veneti* (published in Venice in the 1690s) that Corfu is "Porta di questo Mare, Antemurale d' Italia and Propognacolo della Cristianità ..." ("Gateway to this Sea, Bulwark of Italy and Bastion of Christianity ..."). At the time, certain coastal areas of the Adriatic were under the Habsburg Monarchy (a portion of Istria, the Kvarner coast, and the Velebit region) and the Ottoman Empire (a part of Dalmatia, the territory of present-day Montenegro, and present-day Albania), whereas some smaller parts of southern Dalmatia belonged to the Republic of Ragusa (Dubrovnik). On its western coast, the territories belonged to the Papal States and the Kingdom of Naples, later known as the Kingdom of the Two Sicilies (Chaline et al. 2001; Ivetic 2019).

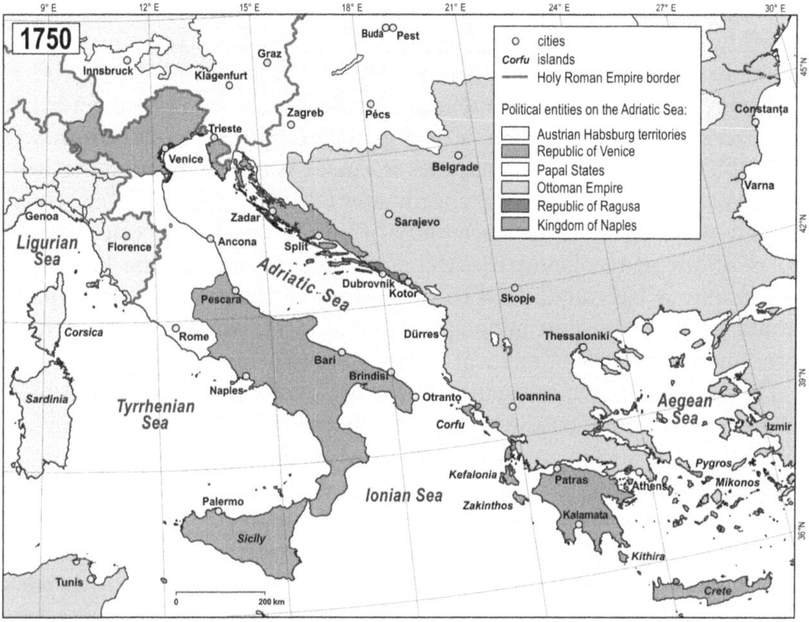

Fig. 1.2 Political map of South and Southeast Europe for the year 1750. Basemap shapefile source: marineregions.org (Claus et al. 2017)

Throughout history, seafaring has been a crucial catalyst for economic and cultural activities and developmental interactions in the Mediterranean (Abulafia 2011), encompassing a complex array of elements such as navigation, shipbuilding, maritime affairs, port management, trade, fisheries, and even warfare. Consequently, seafaring relies on the integration of knowledge and technologies from diverse scientific and professional domains, with nautical cartography holding a particularly significant position. Contemporary nautical charts, predominantly in digital format and integral to electronic navigation systems, are the culmination of a long-standing tradition of maritime connections with various scientific fields, evolving alongside technological advancements. As a result, they constitute an essential part of maritime culture, contributing as a vital aspect of both the tangible and intangible cultural heritage of the Mediterranean regions.

Since the beginning of seafaring, sailors relied on accurate information regarding the geographic characteristics of the maritime regions they traversed to establish safe and precise courses, as maritime navigation is inherently less predictable and more perilous than terrestrial travel. A more accurate understanding of the spatial composition of surrounding geographical features, bearings, and distances between them, along with the subsequent use of spherical coordinates to define their locations, could partially mitigate the unpredictability of sailing and help create nautical charts of adequate planimetric accuracy. This means that the history of cartography is basically linked to the history of the sciences, particularly mathematics, astronomy, geography, and geodesy, which is easily seen by observing the history of nautical cartography (Turnbull 1996).

The earliest known navigational aids are *periploi*, written accounts of sailing routes along coastlines, with distances usually conveyed in days of sailing, created during antiquity, with some of the oldest known examples being the *Periplous of Pseudo-Scylax* (fourth century BCE) and *Stadiasmus Maris Magni* (c. third century) (Taylor 1957; Shipley 2011). During classical antiquity, monumental astronomical, geodetic, and cartographic achievements were undertaken, resulting in assessments of the Earth's size, determining its surface within spherical geometry by assigning latitudes and longitudes to locations, and developing map projections to flatten it into a plane with defined mathematical parameters (Keuning 1955; Dilke 1985; Snyder 1993). However, there is no positive historical evidence that these scientific advancements were employed to create nautical charts during classical and late antiquity.

As far as is known, the roots of nautical cartography date to the late thirteenth and early fourteenth centuries, marked by the sudden historical emergence of *portolan charts* in the Mediterranean (Fig. 1.3). These manuscript charts, drawn with ink on sheets of vellum and primarily focused on the coastline areas, are recognisable for their distinct visual characteristics, which clearly distinguish them from any other known type of earlier-made map. Portolan charts were named after their content-wise similarities to *portolans* (*portolani*), late medieval and early modern texts that contained useful navigational data such as bearings and distances between the significant ports, capes, or river mouths (Campbell 1987; Monmonier 2004; Pujades i Battaler 2007), although certain scholars prefer to label them only as *nautical charts* (Gautier Dalché 1995, 2002). Their most outstanding feature, the exceptionally realistic portrayal of coastlines, is intriguing due to a complete lack of their development path, since the

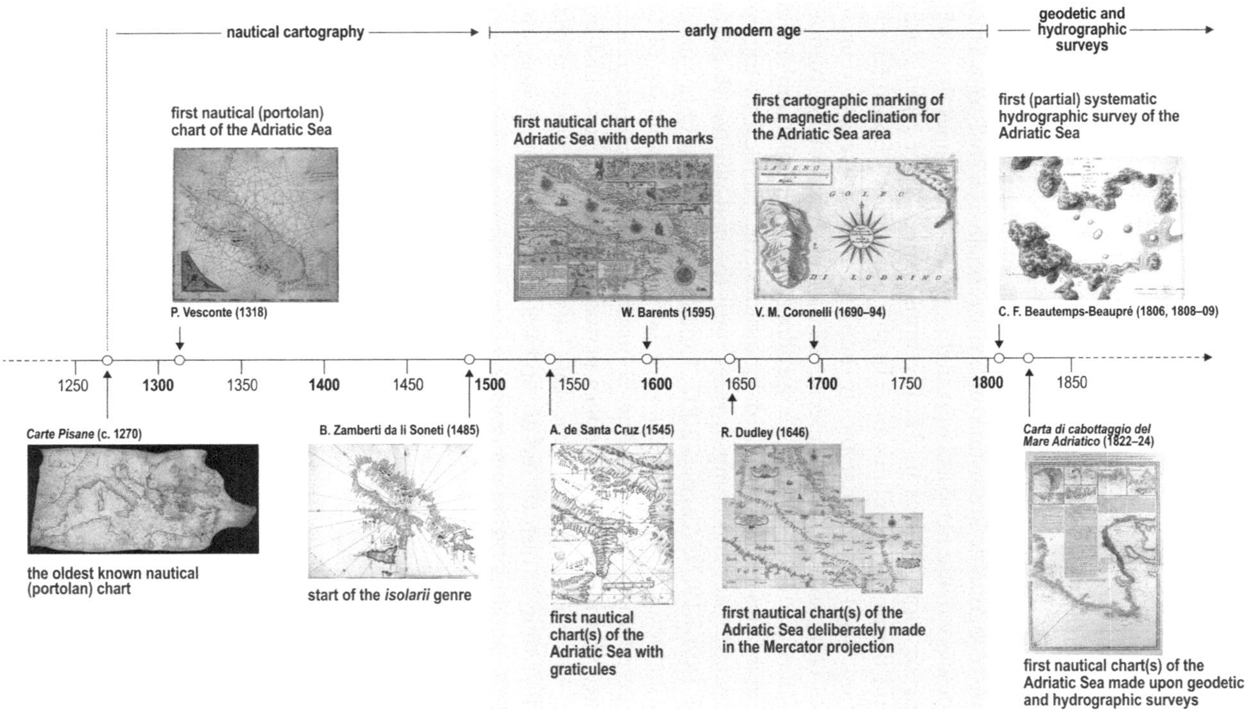

Fig. 1.3 Milestones in the history of nautical cartography for the Adriatic Sea area

earliest known charts, such as the *Carte Pisane*, and the *Cortona chart*, already achieved high levels of planimetric accuracy, strongly resembling a modern map in the Mercator projection. Simultaneously, they also raise questions about their origin because the cartographers who drew them left us no clear information on how the spatial data underlying their geometry was initially observed and assembled into a coherent framework (Nicolai 2014, 2024; Marelić 2024b, 2024c, 2025a). Apart from their geometry, there are other prominent features specific to portolan charts. The most recognisable are wind roses, occasionally referred to as rhumb lines. These are dense networks of straight, colour-coded lines that radiate from one or more centres in 32 directions, named after eight cardinal winds and their cross-combinations. In the later stages of portolan chart production, wind roses were typically supplemented with artistic representations of compass roses, which were occasionally very ornate. Although early modern cartography saw the introduction and gradual implementation of graticules of latitudes and longitudes, the nautical cartography retained wind roses and compass roses until the mid-nineteenth century (Wagner 1896/1969; Astengo 2007; Faričić et al. 2023a). Other notable features include the use of linear scale bars and the inscription of toponyms perpendicular to coastline contours, which were progressively abandoned during the early modern period and became inscribed horizontally. As far as is known, linear scale bars appeared on portolan charts for the first time in history and are still frequently employed, not just in nautical cartography but in general.

Previous research on medieval and early modern nautical charts of the Mediterranean Sea has revealed that only a small number of them were actually used for navigation, as the preserved examples contain no traces of sea salt, punctures made by the usage of compasses (navigational dividers), or other signs of wear typical for the navigational application. Instead, most surviving charts seem to have been made for land-based users interested in the geographical features of the sea and coast for economic, political, or academic purposes (Campbell 1987; Sheehan 2012). Their production placed a great emphasis on aesthetics, leading Corradino Astengo (2007) to classify them as *decorative charts*. These charts are, in many instances, preserved and stored in archives and libraries, unlike those used in navigation, which are now presumed lost, seemingly due to their deterioration caused by shipborne use (Campbell 1987; Astengo 2007). Notably, numerous historical documents confirm that nautical charts were used for navigation, including records indicating that they were a mandatory part of ship equipment (Pujades i Battaler 2007), and one of the most important such historical testimonies is the *De Navigatione Liber* treatise, written in 1464 by Benedetto Cotrugli of Ragusan origin, in which he thoroughly describes how nautical charts are to be used in navigation (Cotrugli 1464). Although the exactitude and verisimilitude of charted areas and the navigational utilisation have always been the most important aspects of nautical cartography, they were often intertwined with elements that are non-navigational in essence, especially during the early modern period. Apart from technological advancements and scientific accomplishments of the time, early modern nautical charts also mirrored the illustrative abilities, expressive styles, geographical expertise, or ideological and political perspectives of their authors. For centuries, cartographers have balanced both scientific precision and artistic expression (Ribeiro and Caquard 2018), a duality that is evident in the interplay between art and science within cartography that encompassed both linear and "painterly" techniques (Karssen 1980; Krygier 1995; Cartwright 2009), and some historians of cartography assert that a map "can be observed like other artistic phenomena: creation of objects, medium of composition, circulation of motifs, and their evolution" (Vagnon 2007: 293). However, the question arises as to how a nautical chart could have concretely served sailors and other interested users in this context. To nautical chart users, cartographical representations suggested different messages and presented various perspectives on the depicted maritime, island, and coastal spaces. In general, maps are more interpretations of facts than objective representations of facts that are devoid of any given meaning. Maps contain ideological and rhetorical devices that are intertwined with depictions of geographical objects in physical space. Therefore, different types of maps "can document a social history of power, especially power over space" (Wintle 1999: 137). In a similar manner, Emanuelle Vagnon highlights that: "Mapping the seas is a way of knowing, discovering, and conquering maritime space" (Vagnon 2021: 170).

Some of the greatest scientific and technological challenges in human history—such as methods on how to effectively plot the rhumb lines (loxodromes) as straight lines on a map accomplished by the Mercator projection in the late sixteenth century, the observations and mappings of the magnetic declination on a global scale since the eighteenth century, and the possibility to accurately observe the longitudes on the open sea accomplished by the invention and the use of the chronometer during the eighteenth century—were often either triggered or promptly adopted by mariners and nautical cartography (Taylor 1957; May 1973; Hewson 1983). The full integration of science and cartography was achieved in the era of the first geodetic surveys (Turnbull 1996). Since the mid-eighteenth century, the procedure of acquiring and processing spatial data and visualising them on maps became institutionalised and based on the precisely established geodetic surveying methods in order to produce the official topographic maps. Soon after, the hydrographic surveys were employed, whose spatial data was used to produce modern and highly accurate nautical charts (Fig. 1.3). The application and development of contemporary technologies, such as radio and satellite positioning, high-resolution seabed surveying through echo sounding, and computer technology adoption, led to further advancements in the field during the second half of the twentieth and twenty-first centuries. As a result, nautical charts are now a highly accurate and strictly standardised branch of cartography that utilises all the advantages of the geographic information system (GIS) environment.

1.1 The Advent of Nautical Cartography and Mathematical Approaches to Chart-Making

The Adriatic Sea appeared on nautical charts showing the entire Mediterranean, only the central or eastern Mediterranean, as well as the standalone geographical area covering the majority of the map field. In *isolarios* (or *isolarii*) and nautical atlases, some of their smaller segments, especially the ports and harbours, were mapped to a larger scale and shown in more detail. Apart from the anonymous *Carte Pisane*, presumably created in the late thirteenth century (c. 1270), on which the Adriatic Sea is shown together with the rest of the Mediterranean, the first nautical (portolan) chart on which the depiction of the Adriatic Sea covers the entire field was created by Pietro Vesconte in Venice in 1318 (Figs. 1.3 and 1.4).

Vesconte's creation was followed by numerous other Venetian cartographers, for example, Francesco de Cesanis (1421), Giacomo Giroldi (1426), Grazioso Benincasa (1472), and Bartolommeo Zamberti da li Sonetti (1485). Their charts were produced either as individual cartographic works or as parts of comprehensive publications. Bartolomeo Zamberti da li Sonetti, for example, published his nautical chart of the Adriatic Sea (Figs. 1.3 and 1.5) in *Isolario*, one of the prototypes of this genre—a kind of compendium of maps and descriptions of individual islands and harbours that was published from the fifteenth to the seventeenth century (Tolias 2007; Stouraiti 2013).

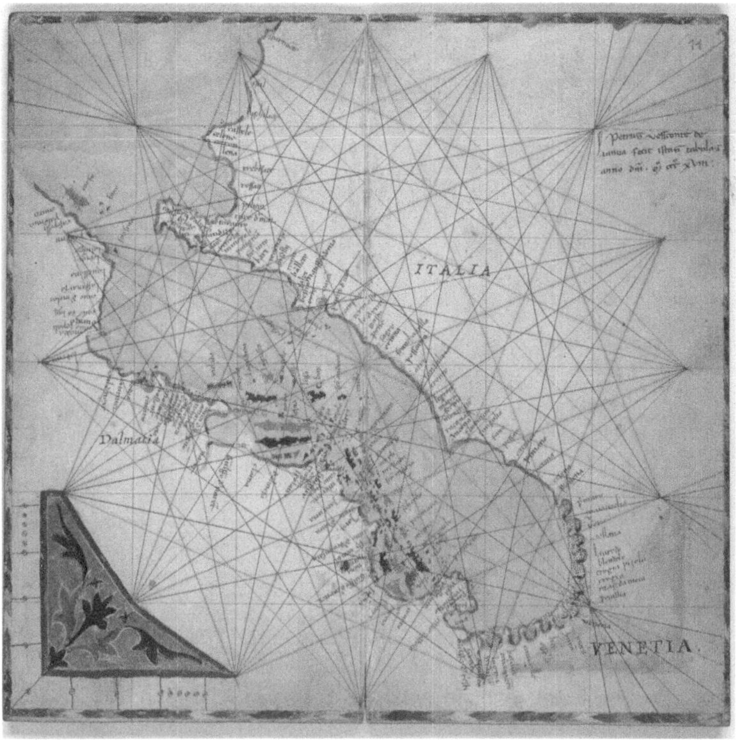

Fig. 1.4 Pietro Vesconte, nautical chart of the Adriatic Sea, Venice, 1318—the first manuscript nautical chart of the Adriatic Sea (Source: Austrian National Library, Vienna, Call Number: Cod. 594 (Cimel. 20), 10v-11r)

During the sixteenth century, the established medieval practice of producing manuscript nautical charts of the Adriatic Sea continued. In the mid-sixteenth century, woodcut printing began to be used (the first such nautical charts were printed by Pietro Coppo in 1525 and by Giovanni Andrea Vavassore in 1539), followed by the implementation of the copperplate printing technique. Although printed nautical charts of the Adriatic were produced from the first half of the fifteenth century, manuscript charts continued to be made in parallel until the end of the early modern period. Because printed charts were more widely available due to the possibility to produce numerous identical (paper) copies at drastically lower production costs, they held greater communicative potential.

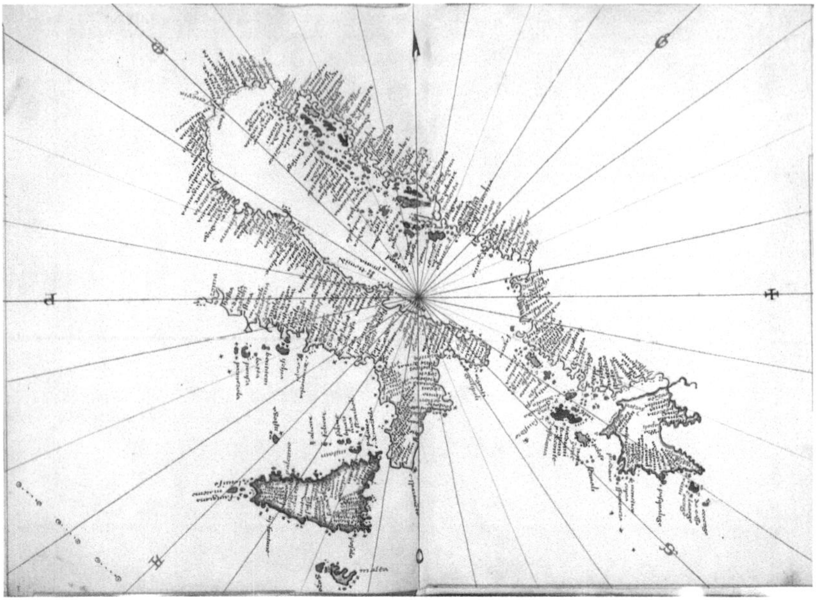

Fig. 1.5 Bartolomeo Zamberti da li Soneti, nautical chart of the Adriatic Sea, Venice, 1485—an example of a chart that is part of comprehensive publications (Source: National Maritime Museum, Greenwich, London; P/21(2); MS 38-9920C)

In parallel, the Adriatic Sea was also depicted on geographical maps, many of which were strongly influenced by its appearance on Claudius Ptolemy's *Fifth Map of Europe*. Although Ptolemy's *Geography*, originally written in the second century CE, was unknown to Western European cartographers until the early Renaissance, when the book was translated from Greek into Latin and subsequently printed in various editions (Kozličić 1995; Berggren and Jones 2000; Gautier Dalché 2007), it exerted significant influence on early modern cartographic representations. Between 1539 and 1560, Alonso de Santa Cruz made *Islario general de todas las islas del mundo*, which contains three charts of the Adriatic (Fig. 1.6), accompanied by their geographical descriptions, many of which were derived from classical antiquity sources (Faričić and Lončar 2024). Unlike other Renaissance authors of comparable works, including Gratiosus Benincasa, Francesco Berlinghieri, and Benedetto Bordone, his charts feature the earliest known representation of latitudes and

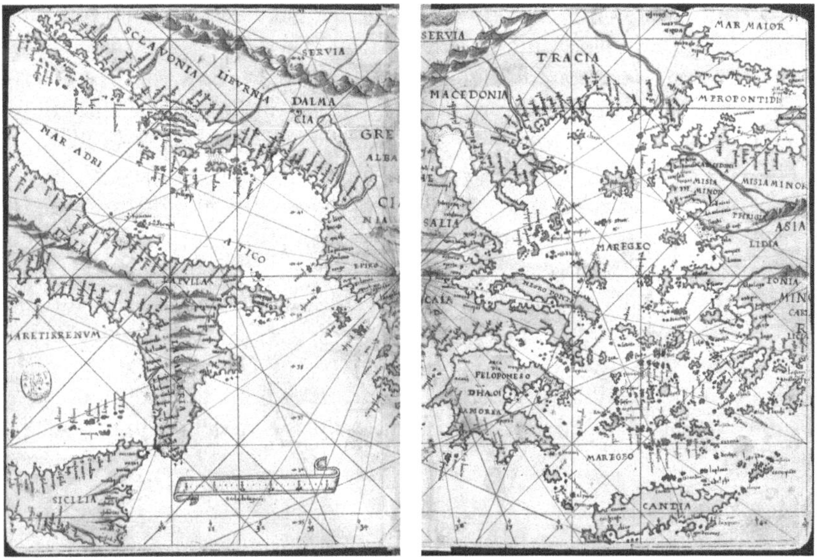

Fig. 1.6 The nautical chart of the northern part of the Central Mediterranean from Alonso de Santa Cruz's *Islario general de todas las islas del mundo* (1539–1560) containing the markings of latitudes and longitudes (Source: National Library of Spain (Biblioteca Nacional de España), Madrid, MSS.MICRO/12638)

longitudes on nautical charts for that region (see Fig. 1.3), with longitudes seemingly approximated based on Ptolemy's geographical information.

1.2 *Stagnation and Uncritical Reproduction of Obsolete Imagery*

During the sixteenth and much of the seventeenth centuries, nautical charts generally offered more accurate depictions of the Adriatic coastline and included a greater number of geographical names for harbours, bays, capes, and islands in comparison with geographical maps. After this period, however, geographical maps began to surpass nautical charts in terms of the quality of their geographical content. During the seventeenth and eighteenth centuries, printed nautical charts of the Adriatic tended to circularly reproduce geographic contents that were already obsolete, mainly due to uncritical compilation of elements that were derived from previous works. Some notable examples are nautical charts published in series by several Italian, Dutch, and English cartographers in the seventeenth century, which were, in fact, almost identical copies of the nautical chart published by the Dutch cartographer and publisher Willem Janszoon Blaeu in Amsterdam in 1621. Interestingly, this highly influential chart by Blaeu was actually a somewhat more generalised and slightly modified version of a nautical chart of the Adriatic Sea made and published in 1595 by Dutch cartographer and explorer Willem Barentsz in Amsterdam (Fig. 1.7). His chart is also the first one that contains sporadic depth-marks of the Adriatic area that are inscribed within the large-scale insets placed in its upper right and bottom left corners (see Fig. 1.3). In 1646, the English cartographer Robert Dudley finished his *Dell'arcano del mare* book in Naples, the first nautical publication accompanied by charts that were deliberately made in

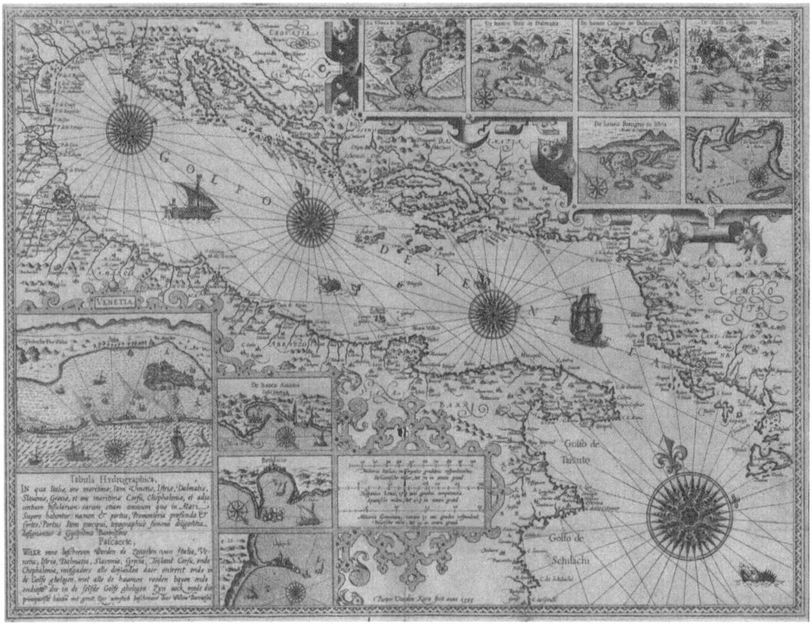

Fig. 1.7 Willem Barentsz, nautical chart of the Adriatic Sea, Amsterdam, 1595—prototype of printed nautical charts of the Adriatic Sea (Source: Stanford University Libraries, The Barry Lawrence Ruderman Map Collection, Stanford)

the Mercator projection, three of which partially depict the Adriatic Sea (see Fig. 1.3). Although his maps included graticules of latitudes and longitudes, Dudley merely overlaid them on Blaeu's image of the Adriatic (Marelić 2023a).

A somewhat different perspective on the maritime and geographical features of the Adriatic Sea was offered by Ottoman cartographers. Among the most significant of these is the sailor and cartographer Muhiddin Pîrî Reis, who published the *Kitâb-ı Bahriye*—a navigational guide containing numerous maps depicting both the Adriatic Sea as a whole (Fig. 1.8) and specific segments of its coast—several times during the first half of the sixteenth century. Pîrî Reis's original contribution is particularly evident in his detailed depictions of Croatian islands and various Adriatic ports, which were notably superior to the cartographic representations appearing in the isolarios made by Venetian and other European authors in the sixteenth and seventeenth centuries (Kozličić et al. 2015; Mlinarić and Faričić 2024).

The largest portion of nautical charts and geographical maps of the Adriatic Sea were made by Venetian cartographers, whereas the

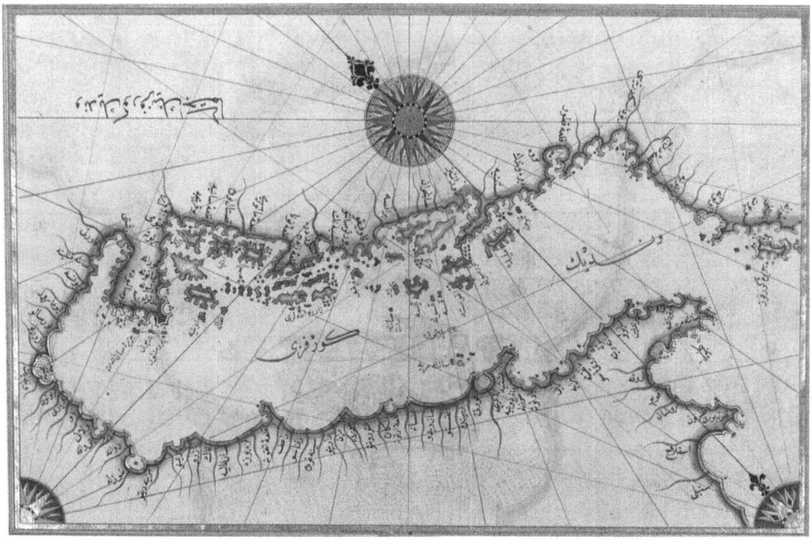

Fig. 1.8 Muhiddin Pîrî Reis, nautical chart of the Adriatic Sea, Gallipoli, 1526 (Source: The Walters Art Museum, Baltimore; W.658, fol. 208a)

cartographers from other parts of Europe often copied their geographical content. Consequently, it is not surprising that Dutch, French, English, and other cartographers did not depict the Adriatic any better than their Venetian predecessors or contemporaries. This is easily confirmed by comparing the shape of the mainland and island coastline contours, and the geographical names found on the Venetian nautical charts with the equivalent elements found on nautical charts made by non-Venetian cartographers. Nevertheless, until the early nineteenth century there were no significant improvements in the accuracy of coastline representation or geographical names, with the exception of the marking of sea depths in more detailed representations of individual Adriatic harbours that have been mapped to a larger scale. For example, the manuscript (portolan) charts of the Adriatic Sea from the sixteenth and seventeenth centuries—made by Battista Agnese (1538), Diogo Homem (1570), Vincus Demetrius Voltius (1593), and Marcheto Fassoi (1679)—hardly differ in their geometry and their geographical content overall (see Fig. 1.9), whose immense geometric resemblance was cartometrically examined in detail (Marelić 2023a, b, 2024a). Although Agnese, Homem, and Fassoi produced their charts in Venice, Agnese was of Genoese origin, whereas Homem came to Venice from Portugal. Voltius made his 1593 chart in Naples, although he also worked in Livorno, and his origins were from the Republic of Ragusa, a city-state that was a political rival of the Republic of Venice at the time. Fassoi's chart of the Adriatic, also made in Venice, was created around a century later and represents a nearly identical copy of its coastline renderings.

Up to the end of the eighteenth century, along with their Venetian counterparts, French cartographers showed the most intense interest in the Adriatic. In their nautical atlases of the Mediterranean and Adriatic Seas, they retained the "Venetian image" of the Adriatic while introducing newly made large-scale renderings of individual Adriatic harbours. Detailed depictions of the larger Adriatic harbours and a general depiction of the entire Adriatic Sea (Fig. 1.10) were published in *Description Géographique du Golphe de Venise et de la Morée*, made by the French cartographer and King's hydrographer Jacques Nicolas Bellin in Paris in 1771, and in *Recueil des Principaux Plans des Ports et Rades de la Mer Mediterranée*, made by the French royal hydrographer Joseph Roux in Marseilles in 1764, whose second edition was published in Genoa in 1779. Jean Joseph Allezard published a revised edition of Roux's work in Genoa in 1804. The nautical charts and plans in these publications did not change the

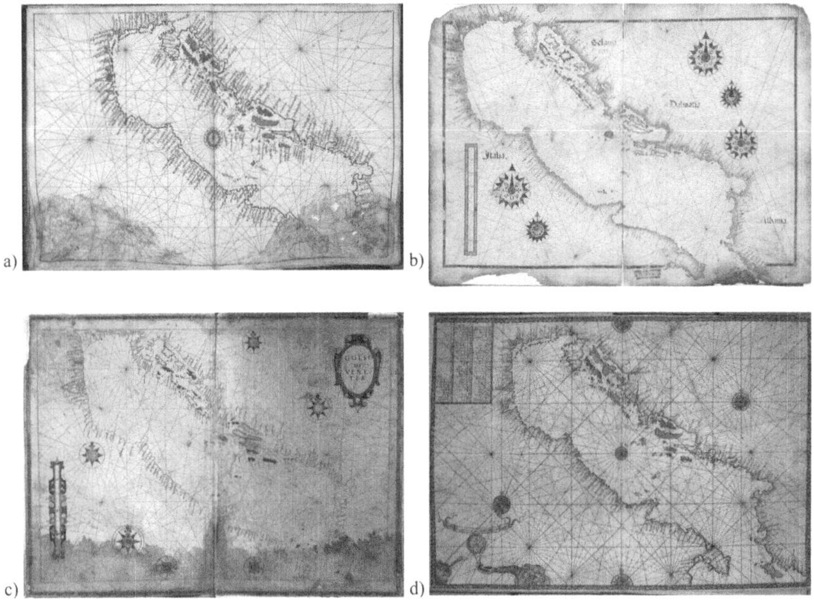

Fig. 1.9 Four portolan charts of the Adriatic Sea with nearly identical coastline renderings: (**a**) Battista Agnese, 1538 (Source: University of Pennsylvania, Kislak Center for Special Collections, Rare Books and Manuscripts, Philadelphia; LJS 28), (**b**) Diogo Homem, 1570 (Source: Croatian State Archives, Cartographic Collection, Zagreb; HR-HDA-902, D.XIV.6), (**c**) Vincus Demetrius Voltius, 1593 (Source: National Library of Finland, Maps, The Nordenskiöld Map Collection, Helsinki; N-Kt-103c), (**d**) Marcheto Fassoi, 1679 (Source: The Huntington Library, Art Museum, and Botanical Gardens, Library Collections, Maps and Atlases, Portolans, San Marino, CA, USA; mssHM 30)

overall cartographic perception of the Adriatic Sea basin and were not able to meet the concurrent, more sophisticated demands for reliable and geodetically well-defined cartographic frameworks. Bellin's chart from 1771 contains one of the earliest known pieces of information about the contemporary values of magnetic declination for the Adriatic Sea (see enlarged circular inset in Fig. 1.10). As far as is known to the authors, the earliest known cartographic representation of magnetic declination for the Adriatic area is V. M. Coronelli's large-scale chart of the Sazan island (*Saseno*) (Figs. 1.3 and 1.11), published in his Isolario *Mari, Golfi, Isole, Spiaggie,*

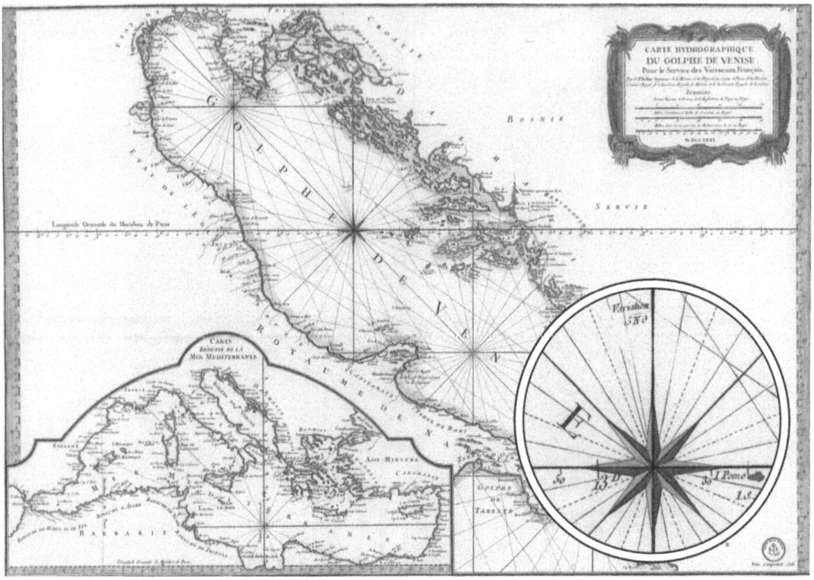

Fig. 1.10 Jacques-Nicolas Bellin, *Carte hydrographique du Golphe de Venise*, Paris, 1771, containing magnetic declination data for the Adriatic Sea area; see enlarged circular inset (Source: National and University Library, Map and Atlas Collection, Zagreb; S-JZ-XVIII-145)

Porti, Citta, Fortezze, Ed altri Luoghi Dell'Istria, Quarner, Dalmazia Albania Epiro, e Livadia, Delineati, e Descritti Dal P. Generale Coroneli (1690–94) (Marelić and Faričić 2023).

1.3 Demands for a More Accurate Representation

Due to the fact that no systematic geodetic or hydrographic surveys were conducted in the Adriatic region before the early nineteenth century, its cartographic representations remained notably anachronistic, retaining a style and geographic content that had changed little since late medieval times. Even the nautical charts of the Adriatic that were produced in the late eighteenth and early nineteenth centuries and published with pretentious claims to be "new" (*nuova carta marittima*) did not make any significant contributions to the development of nautical cartography. Some more renowned examples of such practice are charts published by Ludovico

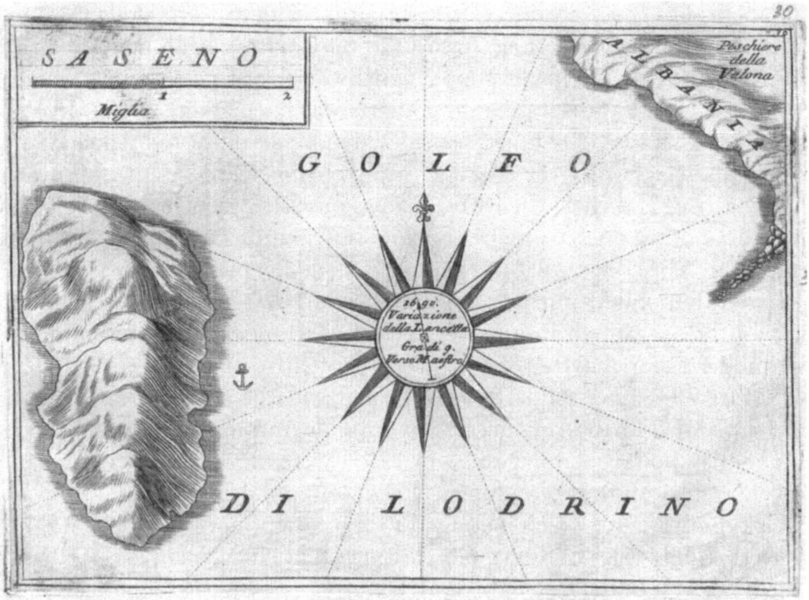

Fig. 1.11 Vincenzo Maria Coronelli's nautical chart (1690–1694) of the Sazan island (Chart source: Vincenzo Maria Coronelli, Isolario, Vol. IV, Venice, 1690–1694 (Source: State Archives in Zadar, Library, Zadar; II.A*)

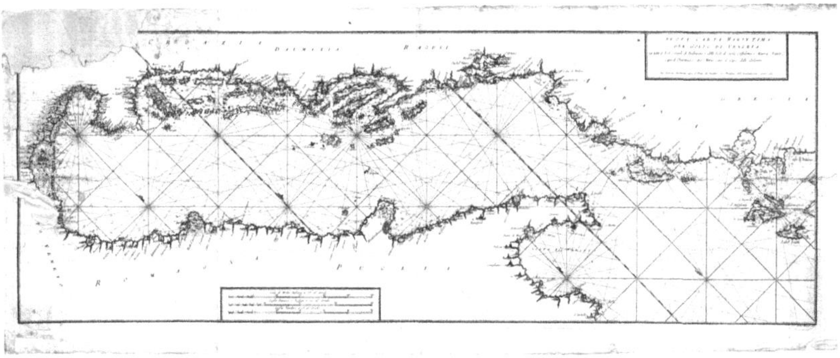

Fig. 1.12 Lodovico Furlanetto, *Nuova carta Marittima del Golfo di Venezia*, Venice, 1784—a medieval nautical cartographic image of the Adriatic Sea in a printed edition from the end of the early modern period (Source: State Archives in Zadar, Cartographic collection, Zadar; HR-DAZD-383 No. 3.1)

Furlanetto in 1784, Vincenzo De Lucio from 1794 to 1796, and Giovanni Grubas in 1809 (Rossit et al. 2006). On Furlanetto's 1784 nautical chart (Fig. 1.12), the only major supplement in terms of geographic content compared to late medieval nautical charts was his marking of sea depths. From 1794 to 1796 De Lucio provided a more accurate and detailed cartographic depiction of the Adriatic (in nineteen sheets) in comparison to Furlanetto's 1784 chart. In 1809, Grubas published new nautical chart of the Adriatic Sea (based on De Lucio's observations and his charts). In spite of its self-congratulatory aspects, it contains many errors that were later rightly pointed out by the English hydrographer and cartographer William Henry Smyth in his overview of charting the Mediterranean in the modern era (Smyth 1854). In terms of their quality in comparison with the 1784, 1794, and 1809 nautical charts, the charts in the *Atlante Marittimo delle due Sicile*, made by Giovanni Antonio Bartolomeo Rizzi Zannoni and published in Naples in 1792, were more accurate. Pursuant to field observations he conducted, Rizzi Zannoni depicted the southwest side of the Adriatic, which belonged to the Kingdom of the Two Sicilies, in great detail (Valerio 2006). His cartographic achievement is considered one of the best of the late eighteenth century (Tooley 1962). However, its content does not encompass the entire Adriatic area.

The turning point in Adriatic cartography, following earlier partial, though scientifically based surveys during the French administration, was mostly the result of work led by Ferdinando Visconti and Charles François Beautemps-Beaupré in 1806 and 1808–1809 (Kozličić 2006; Valerio 2006; Faričić and Mirošević 2017) and was denoted by the first systematic hydrographic survey, which covered the entire Adriatic Sea and resulted in the publication of the *Carta di cabottaggio del Mare Adriatico* in Milan in 1824 (see Fig. 1.3). This achievement was the result of cooperation between institutions in three states—the Austrian Empire, the Kingdom of the Two Sicilies (Kingdom of Naples), and the United Kingdom of Great Britain and Ireland—and was the first international scientific campaign in the field of cartography. The survey was coordinated by Antonio Campana, Director of the Military-Geographic Institute in Milan (Marieni 1830; Smyth 1854; Valerio 2006). Captain Smyth, on behalf of the Hydrographic Office of the Admiralty of the United Kingdom of Great Britain, in his review of preparations for a hydrographic survey of the Adriatic Sea, cited Captain Thomas H. Hurd, who had said during a presentation at the Admiralty: "All the charts of the Adriatic that I have seen are erroneous in regard to its eastern shores... I would therefore recommend a survey to be undertaken" (Smyth 1854: 364). One of the participants in the surveys of the Adriatic during 1818 and 1819 was Giacomo Marieni, who likewise could not understand the lack of accurate charts of the Adriatic. In the *Portolano del Mare Adriatico* published by the Military-Geographic Institute in Milan in 1830, Marieni wrote: "The Venetians, who along with their well-known success, have dominated the Adriatic for several centuries, for some reason, whether because the times were like that, or perhaps because of the doubtful character of the Republic, which is not for us to judge, never considered producing nautical charts of their shores, which are so desired by all sailors who have gained experience in other seas and know they can make sailing more pleasant, safe, and quick" (Marieni 1830: VI).[1] Hurd and Marieni's statements indicate there was stagnation in terms of developing the nautical cartography of the Adriatic Sea during the eighteenth century. The publication of the *Carta di cabottaggio del Mare Adriatico* allowed sailors to resolve various navigational tasks for the first time using highly accurate nautical charts, marking the demise of early modern nautical cartography, as nautical charts have since been produced by geodetic and hydrographic state offices rather than small, often family-run workshops based on their directors' authority (Faričić 2011; Mlinarić and Miletić Drder 2017).

2 The Spectrum of Scholarly Approaches and Methods

Nautical charts played a pivotal role in the early modern period, which explains their prominence as research subjects across various scholarly fields, particularly geography, cartography, and history. The available academic literature demonstrates that numerous studies have focused on early modern nautical charts of the Adriatic Sea, thereby advancing our understanding of both their production processes and the geographical information they convey. In the history of cartography-related monographs, these charts are generally examined as part of the broader body of old maps depicting the Adriatic, regardless of their specific type or origin.

[1] Marieni's original statement in Italian: "I Veneziani, i quali dominarano d' Adriatico per alcuni secoli, e con quella fortuna che ognun sa, qualunque ne fosse la causa, o la condizione de' tempi, o forse l' indole sospettosa della repubblica, chè a noi non septta do far simili ricerche, non hanno mai pensato alla Carta nautica delle loro coste, tanto desiderata, e massime da que' navigatori che già sapevano per esperienza fatta in altri mari quanto avrebbe giorato al comodo, alla sicurezza de' viaggi, e a renderli più spediti" (Marieni 1830: VI).

2.1 Studies Related to the Adriatic Area

Old nautical charts can be studied from various standpoints and with various aims, examining their extremely varied communication capacities, particularly as part of recent cartographic discourse regarding multiple readings and interpreting the symbolic strength of extra-contextual communication (Panofsky 1970; Mlinarić and Miletić Drder 2017). In general, maps are rich, multi-layered sources of information about the depicted areas, the methods used to create them, their authors, and the circumstances of their production (Bagrow and Skelton 1985). In addition, by applying the principle of analogy to iconographic or imagologic interpretation, including decomposition or deconstruction (Harley 1988, 1989; Monmonier 1996; Dukić et al. 2009; Mlinarić and Gregurović 2011; Mlinarić 2012; Šakaja 2015), it is possible to read old maps in terms of their symbols (Fürst-Bjeliš and Zupanc 2007). Starting from research achievements so far and the discourse tenets of a "new cultural cartography" (Harley 1988, 1989), and pursuant to the interdisciplinary approach that acknowledges geographic, historiographic, geodetic, and nautical discourses, it is necessary to supplement the existing knowledge of early modern nautical charts of the Adriatic Sea, particularly in terms of applying cartometric analyses and evaluating cartographic sources as media of communicating in space and about space.

One of the earliest such works was *Alte handschriftliche Schifferkarten in den Bibliotheken zu Venedig* by Petar Matković, written in 1862, in which he described manuscript nautical charts in the Venetian archives and libraries (Matković 1862). Adolf Erik Nordenskiöld used Matković's work as a source in *Periplus: An Essay on the Early History of Charts and Sailing-Directions* in 1897 (Nordenskiöld 1897). This is a monumental work, a synthesis of all the author's known information on links between seafaring and nautical cartography, beginning in classical times and ending in early modern geographic discoveries by European seafarers in the polar areas and Pacific Ocean. Although the book does not dedicate a separate chapter to the Adriatic Sea and its depiction on nautical charts, it is mentioned several times in the wider context of nautical issues. The book also contains 60 tables and reproductions of numerous nautical charts, many of which depict the Adriatic, either separately or as part of a larger spatial extent. Other significant contributions at that time, works on old nautical charts that appeared in the late nineteenth and early twentieth centuries, include *Sammlung mittelalterlicher Welt-und Seekarten italienischen*

Ursprungs und aus italienischen Bibliotheken und Archiven by Theobald Fischer (1886), *Die italienischen Portolane des Mittelalters. Ein Beitrag zur Geschichte der Kartographie und Nautik* by Konrad Kretschmer (1909), and *Portolan Charts—Their Origin and Characteristics with a Descriptive List of Those Belonging to The Hispanic Society of America* by Edward Luther Stevenson (1911). Relatively newer works of this kind include chapters from the book series *The History of Cartography* published by the University of Chicago Press. Chapters in this series that are most closely related to the nautical cartography of the Late Middle Ages and early modern period are *Portolan Charts from the Late Thirteenth Century to 1500* by Tony Campbell (1987), *The Renaissance Chart Tradition in the Mediterranean* by Corradino Astengo (2007), and *Maps and Exploration in the Sixteenth and Early Seventeenth Centuries* by Felipe Fernández-Armesto (2007). Among more recent syntheses, *Rhumb Lines and Map Wars: A Social History of the Mercator Projection* by Mark Monmonier (2004) and *Les cartes portolanes: la representació medieval d'una mar solcada* by Ramon J. Pujades i Bataller (2007) are prominent.

More detailed overviews of the development of nautical charts of the Adriatic Sea can be found in *Imago Adriae* by Luciano Lago (1998), *Imago Italiae* by L. Lago et al. (2002), *Coste del Mediterraneo nella cartografia europea: 1500–1900* by Paola Presciuttini (2004), *Descriptio Croatiae* by Mirko Marković (1993), and *Kartografski spomenici hrvatskoga Jadrana* by Mithad Kozličić (1995), along with a book entitled *Five Centuries of Maps and Charts of Croatia* edited by Drago Novak, Miljenko Lapaine, and Dubravka Minarić (2005), the monograph *Povijesna kartografija: kartografski izvori u povijesnim znanostima* by Mirela (Slukan) Altić (2003), and the monograph *Zbirka Novak: Mappae Croaticae u Zbirci zemljovida i atlasa NSK* by D. Mlinarić and Mira Miletić Drder (2017).

Many academic papers have been published on nautical charts of the Adriatic Sea, or on particular aspects of such charts. Italian scholars, such as Roberto Almagia, wrote many prominent works, including one on the nautical charts in the Apostolic Library in the Vatican (Almagia 1944). L. Lago and Claudio Rossit have dealt exhaustively with the nautical cartography opus of Pietro Coppo (Lago and Rossit 1984), while Rossit and others produced an overview of depictions of the Adriatic on old maps, including some nautical charts (Rossit et al. 2006). Vladimiro Valerio published several works with the results of his research into charting the Adriatic, some of which were the result of activities in which the Kingdom of Naples participated (Valerio 2006), and Orietta Selva researched Venetian cartography as a symbol of power and means of planning (2013). Kozličić has made some of the greatest contributions in the field of nautical cartography of the Adriatic. His preoccupation is with contextualising nautical charts as sources from which the developmental trajectory of geographic and oceanographic knowledge of the Adriatic can be traced (Kozličić 1999; Kozličić and Faričić 2004, 2016), and in particular, the hydrographic survey of the eastern Adriatic coast conducted in two separate campaigns under the leadership of Charles François Beautemps-Beaupré (Kozličić 2006). Kozličić has also correlated nautical charts and sailing manuals, which together form a functional whole (Kozličić et al. 2012; Kozličić 2013; Kozličić and Uglešić 2015). Dušanka Čanković has given a succinct historiographic overview of the development of the nautical cartography of the Adriatic from the late fifteenth to the early nineteenth century (Čanković 1987, 2005). The works of M. (Slukan) Altić are particularly valuable, such as those on the development of portolan charts (Slukan Altić 1998, 2003) and on the British, Austro-Hungarian, and Italian contributions to the development of the nautical cartography of the Adriatic (Slukan Altić 2016; Altić 2021). More recently, Josip Faričić, Lena Mirošević, Ivka Kljajić, and D. Mlinarić examined the use of compass roses on nautical charts of the Adriatic Sea (Faričić et al. 2023a), whereas J. Faričić, O. Selva, and Dragan Umek studied geographical names of the Adriatic Sea on old maps and nautical charts (Faričić et al. 2023b). These works are dominated by the descriptive historiographic approach and a qualitative analysis of various spatial data, particularly toponyms, but they fail to include a quantitative cartometric analysis of nautical charts, in the sense of establishing their mathematical and geodetical basis (particularly map projections and scales), which is key to understanding their applicability as reliable sources of information to be used for planning and implementing various navigational tasks. Tome Marelić's doctoral thesis (Marelić 2020), published as three articles (Marelić 2022a, b, 2024a), is the sole extensive and detailed cartometric study of late medieval and early modern nautical charts for the Adriatic Sea in existence. Additional information on cartometric analyses and notable works can be found in a separate section of this chapter.

Croatian historians of cartography have so far paid little attention to reading old nautical charts as sources for establishing concepts of information organisation and imagining the Adriatic Sea as an area of multiple socioeconomic and particularly political and cultural permeation, but this has been done, at least for individual parts of Croatia, in research into early modern geographic maps (Fürst-Bjeliš 2009, 2012; Mlinarić 2012, 2014; Mlinarić and Gregurović 2011). In fact, up to the sixteenth century, Croatia was not shown on large-scale maps, which was not unusual. It was only in the late fifteenth century and early sixteenth century that maps of the region of European states began to be produced, including Croatia and the neighbouring areas of modern Italy and Hungary (Bagrow and Skelton 1985). This initiative in European cartography was prompted on the one hand by the publication of Ptolemy's *Geographia*, with the development of science and the aim of acquiring and depicting new geographic spatial knowledge, and the needs of states and lower levels of government to visualise their territories or the spatial resources at their disposal on the other (Woodward 2007). In terms of the cartographic depiction of individual regional units on the eastern Adriatic coast, we should add the increased interest of European political and cultural centres in understanding the arenas of conflict between the imperial powers of Venice, the Habsburg Monarchy, and the Ottoman Empire (Fürst-Bjeliš and Zupanc 2007; Mlinarić 2014; Mirošević and Faričić 2015). These powers confronted each other with varied outcomes, attempting by different means, including maps, to present their agendas or at least their perceptions of their borders, in order to achieve their political and economic strategies (Mlinarić and Gregurović 2011). During the entire early modern period, this left a strong imprint on the landscape and identity of the Dalmatian coastal area, the contact point for the intermingling of various ethnic, religious, and linguistic structures (Fürst-Bjeliš 2009, 2012). It was logical to expect that nautical charts would also become a medium for transmitting various political messages and not merely navigational tools in the hands of sailors.

2.2 *Qualitative Approaches to the History of Nautical Cartography*

In the contemporary research discourse on reading cartographic sources and their trans-media and trans-genre comparison with other types of visual and narrative historical documents, maps are seen critically as subjective, authorial products that may, but need not, reflect historical truths, at least not generally accepted ones (Brković and Mlinarić 2013). In any case, they reflect the attempt to communicate ideas and concepts (geographic, cultural, geostrategic, in relation to political propaganda, economic, linguistic, ethnic, demographic, and so on) by the people who commissioned them or financed them, and, indirectly, by the authors themselves. In this context, nautical charts were a sophisticated means of expressing the mutual relationships and aspirations of the European political elite, conveyed in coded messages. At the same time, as sources, they imply a certain level of speculation as to their historical witness and are of dubious reliability, although due to their very existence, they are, in a sense, exact, credible historical sources, even when their credibility relates primarily to their intention or aim (Mlinarić and Miletić Drder 2017).

The subjectivity within the mapmaking, which was profound before the implementation of systematic field surveys, was conditioned not only by cartographers' professional competence and their personal preferences but also by the particular purpose of the map and the motifs and interests of the person or institution who commissioned its creation. The purpose of charts and maps was not only to present the best possible depiction of geographical reality but also to shape the reality in accordance with certain political aspirations and economic interests. This can be seen, for example, in the drawing of political boundaries (deliberately anachronistic in terms of changes and new situations that were not accepted for state or legal reasons), the use of geographic names, etc. For example, the reasons behind the Venetian cartographer's decision to write *Golfo di Venezia* for the entire Adriatic Sea are fully internal in terms of contemporary politics, whereas the names *Golfe de Venise* and *Gulph of Venice*, used by French and English cartographers, are adoptions of externally established political narratives. The Venetian cartographers were, thus, emphasising and promoting the current territorial aspirations of the Republic of Venice, whereas the syntagm "Adriatic Sea" was understood as its historical name from classical antiquity (e.g. on his 1688 map of the Adriatic Sea and the surrounding areas, V. M. Coronelli wrote: *Golfo di Venezia olim Adriaticum Mare*, meaning *The Gulf of Venice once upon a time the Adriatic Sea*). The French, English, and other Western European cartographers adopted the Venetian name mostly because they used charts and maps made by the Venetian authors as templates. At the same time, the Austrian cartographers, who refused to acknowledge the dominance of Venice in the Adriatic, called it by its older, politically neutral name—*Mare Hadriaticum*, or *Mare Adriaticum* (Ger. *Adriatisches Meer*, It. *Mare Adriatico*) (Faričić et al. 2023b).

The geographical content of early modern nautical charts of the Adriatic Sea should be studied with regard to the perspective of how geographical reality was observed and cartographically visualised. The working hypothesis is that the geographic content of nautical charts was conditioned by the amount and quality of the geographical knowledge and from the estimated motifs and reasons for its creation. Therefore, distinct perspectives (of their authors) yield different perceptions of the (same) reality (Todorova 2015), whereas modern epistemology, in an era of different spatial (Space Turn) and similar interpretative reversals that critically examine but also build on Fernand Braudel's concepts of endurance (Braudel 1972), defines spaces as a cultural triad. It simultaneously comprises actual (physical space), intellectually processed mental space, and, finally, the living "third-space." It is characterised by various symbolic meanings, ranging across the material, hybrid, and the representative, particularly when depicted on maps as visual historical sources (Cosgrove 2004; Crampton 2009; Soja 2001).

The quality and range of geographic content on old nautical charts should be considered within the context of the socioeconomic settings of the time, which affected the competence, perspectives, and motifs of their authors. It is logical to assume that cartographers' decisions regarding which geographical features are going to be drawn on the map depended on their spatial relevance—both in terms of their actual size and in terms of their perceived importance (political, navigational, etc.), which do not always coincide. While the physical dimensions, such as lengths or areas, are relatively easy to understand, observe, and even survey, the importance is a non-quantifiable subjective category that varies on an individual basis, especially between those who assess it internally or externally. Rob Kitchin and Martin Dodge have stated quite correctly that "maps are transitory and fleeting, being contingent, relational, and context-dependent" (2007: 335). Modern theories in cartography recognise that maps (whether early modern or modern) provide their authors with liberties to lie about historical and more recent events with ease, despite all the advances of science and human self-assurance (Monmonier 1996; Mlinarić and Gregurović 2011).

By conducting a comparative analysis of early modern nautical charts of the Adriatic Sea and applying a method of deconstruction or "decomposition," this study investigates the perception of the actual geographical space. The findings show that this perception was influenced by the methods used to collect spatial data, the extent and quality of geographical knowledge, and the perspectives from which the Adriatic area was viewed. Different perceptions of the Adriatic reveal diverse ideas, concepts, discourses, and biases (e.g. regarding spatial organisation along core–periphery lines, "Us" vs. "Others," East vs. West, etc.).

On old cartographic representations of the Adriatic, authors or commissioners expressed their intentions by employing suggestive generalisations—highlighting or concealing particular elements of geographical reality—and incorporating iconographic features whose rich symbolism exceeded mere aesthetic (decorative) purposes. The early modern nautical charts of the Adriatic Sea serve as a good example of the interplay between realistic and imaginary content, which intertwines to form complex composites. In these composites, one can discern far more than is revealed by a straightforward interpretation of what and where things are depicted on the charts.

2.3 *Quantitative Approaches to the History of Nautical Cartography*

Cartometric analyses, a relatively new approach and less frequently conducted in comparison to traditional qualitative academic publications, occupy a distinct niche in the history of cartography studies. In the context of old nautical cartography, a certain number of notable works examine late medieval and early modern portolan charts, usually in the form of comprehensive studies, some of which are still ongoing (Loomer 1987; Nicolai 2014, 2015, 2016, 2024; Marelić 2024b, c, 2025a, b).

Cartometric analysis is an umbrella term for various quantitative methods used to estimate the geometric properties of maps (its most plausible map projection, its map scale, etc.). In the late nineteenth century, Hermann Wagner was the first person to implement his method with the aim of establishing the geometric features of medieval portolan charts and coined the term "cartometric method" (Wagner 1896/1969). These methods were further facilitated by the development of information technology in the latter part of the twentieth century, and one of its first adopters was Waldo R. Tobler. In 1966, he utilised a computerised method to assess the geometric properties of old maps and examined Pietro Roselli's 1468 portolan chart (Tobler 1966), whereas in 1977, he proposed the concept of the empirical map projection (Tobler 1977). During the late twentieth century, more intensive development of computer technology expanded the possibilities of cartometric approaches, and one of the most comprehensive studies of the time is Scott A. Loomer's doctoral dissertation *A Cartometric Analysis of Portolan Charts: a Search for Methodology*

(Loomer 1987). Cartometric studies of old charts and maps intensified in the twenty-first century, resulting in an increased number of journal publications (Balletti 2006; Gaspar 2007, 2008, 2013; Gaspar and Leitão 2014; Rajaković et al. 2014; Nicolai 2015, 2018, 2024; Balletti and Gottardi 2015; Livieratos and Boutoura 2018; Marelić 2024a, b, c). These publications range from overviews of theoretical works on the general benefits of georeferencing old maps to exact results achieved by applying cartometric analyses. Some of these works stem from the usage of the *MapAnalyst* cartometric freeware, produced by the Institute of Cartography of ETH Zurich. The software is based on algorithms developed by Dieter Beineke (2001; Jenny 2006; Jenny and Hurni 2011) and provides some basic tools for cartometric analyses, but it is not suitable for examining maps with large longitudinal coverage (Jenny 2010; Nicolai 2018). In contrast, more comprehensive and detailed cartometric studies are carried out employing more advanced GIS and statistical software—none of which is designed specifically for cartometric purposes—allowing the geometry of the old charts and maps to be iteratively compared to the multiple and customisable coordinate reference systems of the modern map.

References

Abulafia D (2011) The Great Sea: a human history of the Mediterranean. Oxford University Press

Almagia R (1944) Carte nautiche e affini dal secolo XIV al XVII esistenti nella Biblioteca Apostolica Vaticana, Monumenta cartografica vaticana, I. Citta del Vaticano

Altić M (2021) Charting newly created statehood: a maritime survey of the Adriatic by the Joint Forces of the Austro-Hungarian and Italian Hydrographic Offices. Cartograph J 58(1):13–28. https://doi.org/10.1080/00087041.2019.1685288

Astengo C (2007) The renaissance chart tradition in the mediterranean. In: Woodward D (ed) The history of cartography, Vol. 3, Part 1—cartography in the European Renaissance. University of Chicago Press, Chicago, pp 174–262

Bagrow L, Skelton RA (1985) History of cartography. Harvard University Press

Balletti C (2006) Georeference in the analysis of the geometric content of early maps. e-Perimetron 1(1):32–42

Balletti C, Gottardi C (2015) Analytical methods for the projective content of Coronelli's central Europe maps. In: Arbore-Popescu G, Dabalà C (eds) The cartography of the Danube and the surrounding areas in and after the times of Vincenzo Maria Coronelli and Luigi Ferdinando Marsili. Corilia, pp 23–33, 87–93

Beineke D (2001) Verfahren zur Genauigkeitsanalyse für Altkarten. Doctoral thesis. Universität der Bundeswehr München

Berggren JL, Jones A (2000) Ptolemy's geography: an annotated translation of the theoretical chapters. Princeton University Press, Princeton

Braudel F (1972) The Mediterranean and the Mediterranean World in the Age of Philip II, vol 1. Harper & Row

Brković I, Mlinarić D (2013) Reprezentacija Dubrovnika u književnosti i kartografiji 17. stoljeća. Književna istorija: časopis za nauku o književnosti 149(45):29–53

Campbell T (1987) Portolan Charts from the Late Thirteenth Century to 1500. In: Harley JB, Woodward D (eds) The history of cartography, Vol. 1—cartography in prehistoric, ancient and Medieval Europe and the Mediterranean. University of Chicago Press, pp 371–463

Cartwright W (2009) Art and cartographic communication. In: Cartwright W, Gartner G, Lehn L (eds) Cartography and art. Springer, pp 9–22

Chaline O, Doumerc B, Ducellier A, Sivignon M (2001) Histoire de l'Adriatique. Seuil

Claus S, De Hauwere N, Vanhoorne B, Souza Dias F, Oset García P, Schepers L, Hernandez F, Mees J (Flanders Marine Institute) (2017). MarineRegions.org

Cosgrove D (2004) Landscape and landschaft. German Hist Instit Bull 35:57–71

Cotrugli B (1464) De Navigatione Liber. Yale University, Beinecke rare book and manuscript library, Call No.: MS 557 (https://collections.library.yale.edu/catalog/2005406)

Crampton JW (2009) Cartography: performative, participatory, political. Progr Hum Geograph 33(6):840–848. https://doi.org/10.1177/0309132508105000

Čanković D (1987) Jadranska obala na zemljopisnim kartama do početka 19. stoljeća u kontekstu razmatranja—karta kao kartografsko i estetsko djelo. Vjesnik bibliotekara Hrvatske 30(1-4):45–56

Čanković D (2005) Five centuries of the Adriatic on maps. In: Novak D, Lapaine M, Mlinarić D (eds) Five centuries of map and charts of Croatia. Školska knjiga, pp 202–224

Dilke OAW (1985) Greek and Roman maps. Cornell University Press, Ithaca, NY

Dukić D, Blažević Z, Plejić Poje L, Brković I (eds) (2009) Kako vidimo strane zemlje: uvod u imagologiju. Srednja Europa

Faričić J (2011) Kartografske aktivnosti u Dalmaciji tijekom francuske uprave 1806.—1813. In: Šanjek F (ed) Hrvati i Ilirske pokrajine. Hrvatska akademija znanosti i umjetnosti, pp 353–377

Faričić J, Mirošević L (2017) Carta di cabotaggio del Mare Adriatico (1822-1824). A turning-point in the development of Adriatic maritime cartography. Imago Mundi 69(1):99–111. https://doi.org/10.1080/03085694.2017.1242846

Faričić J, Kljajić I, Mirošević L, Mlinarić D (2023a) Symbolism of compass roses on early modern nautical charts of the Adriatic Sea. KN-J Cartograph Geograph Inf 73(1):19–37. https://doi.org/10.1007/s42489-022-00129-z

Faričić J, Selva O, Umek D (2023b) Geographical names of the Adriatic Sea on medieval and early-modern maps and nautical charts. J Hist Geograp 82:68–80. https://doi.org/10.1016/j.jhg.2023.09.007

Faričić J, Lončar I (2024) Jadranski otoci u Općem izolaru svih otoka svijeta Alonsa de Santa Cruza. Radovi Zavoda za povijesne znanosti HAZU u Zadru 66:41–81. https://doi.org/10.21857/yl4okf8dx9

Fernández-Armesto F (2007) Maps and exploration in the sixteenth and early seventeenth centuries. In: Woodward D (ed) The history of cartography, Volume 3, Part 1—Cartography in the European Renaissance. University of Chicago Press, pp 738–770

Fischer T (1886) Sammlung mittelalterlicher Welt-und Seekarten italienischen Ursprungs und aus italienischen Bibliotheken und Archiven. Ferdinand Ongania

Fürst-Bjeliš B (2009) Slike i mijene regionalnoga identiteta: geografska imena na kartama ranoga novoga vijeka (odabrani primjeri). In: Skračić V, Faričić J (eds) Geografska imena—Zbornik radova s Prvoga nacionalnog savjetovanja o geografskim imenima. Svečuilište u Zadru and Hrvatsko geografsko društvo, pp 63–68

Fürst-Bjeliš B (2012) Imaging the past: cartography and multicultural realities of Croatian realities. In: Bateira C (ed) Cartography—a tool for spatial analysis. Intech Open, pp 295–312

Fürst-Bjeliš B, Zupanc I (2007) Images of the Croatian Borderlands: selected examples of early modern cartography. Hrvatski geografski glasnik 69(1):5–19. https://doi.org/10.21861/HGG.2007.69.01.01

Gaspar JA (2007) The myth of the square chart. e-Perimetron 2(2):66–79

Gaspar JA (2008) Dead reckoning and magnetic declination: unveiling the mystery of portolan charts. e-Perimetron 3(4):191–203

Gaspar JA (2013) From the portolan chart to the latitude chart. Cartes et Géomatique, Comité Français de Cartographie 216:67–77

Gaspar JA, Leitão H (2014) Globes, Rhumb Tables, and the pre-history of the Mercator projection. Imago Mundi 66(2):180–195. https://doi.org/10.1080/03085694.2014.902580

Gautier Dalché P (1995) Carte marine et portulan au XIIe siècle. Le Liber de Existencia Riverierarum et Forma Maris Nostri Mediterranei (Pise, circa 1200). École Française de Rome

Gautier Dalché P (2002) Cartes marines, représentation du littoral et perception de l'espace au Moyen Âge. Un état de la question. In: Martin JM (ed) Castrum

VII. Zones côtières et plaines littorales dans le monde Méditerranéen au Moyen Age. École français de Rome and Casa de Velázquez, pp 9–33
Gautier Dalché P (2007) The reception of Ptolemy's geography (end of the fourteenth to beginning of the sixteenth century). In: Woodward D (ed) The history of cartography, Volume 3, Part 1—Cartography in the European Renaissance. University of Chicago Press, Chicago, pp 285–364
Harley JB (1988) Maps, knowledge and power. In: Cosgrove D, Daniels S (eds) The iconography of landscape. Cambridge University Press, pp 277–312
Harley JB (1989) Deconstructing the map. Cartographica 26(2):1–20. https://doi.org/10.3138/E635-7827-1757-9T53
Hewson JB (1983) A history of the practice of navigation. Brown, Son & Ferguson
Ivetic E (2019) Storia dell'Adriatico. Un mare e la sua civiltà. In Mulino, Bologna
Jenny B (2006) MapAnalyst—a digital tool for the analysis of the planimetric accuracy of historical maps. e-Perimetron 1(3):239–245
Jenny B (2010) New features in mapAnalyst. e-Perimetron 5(3):176–180
Jenny B, Hurni L (2011) Studying cartographic heritage: analysis and visualization of geometric distortions. Comput Graph 35:402–411. https://doi.org/10.1016/j.cag.2011.01.005
Karssen AJ (1980) The artistic elements in map design. Cartograph J 17(2):121–127. https://doi.org/10.1179/caj.1980.17.2.124
Keuning J (1955) The history of geographical map projections until 1600. Imago Mundi 12(1):1–24. https://doi.org/10.1080/03085695508592085
Kitchin R, Dodge M (2007) Rethinking maps. Progress Hum Geograph 31(3):331–344. https://doi.org/10.1177/0309132507077082
Kozličić M (1995) Kartografski spomenici hrvatskog Jadrana. AGM
Kozličić M (1999) Izmjere dubina istočnog Jadrana tijekom 16.-19. stoljeća. In: Gereš D (ed) 2. hrvatska konferencija o vodama: Hrvatske vode od Jadrana do Dunava. Hrvatske vode, pp 325–334
Kozličić M (2006) The Eastern adriatic in the work of Beautemps-Beaupré. Croatian Hydrographic Institute, Split
Kozličić M (2013) Pomorske karte i plovidbeni priručnici istočnog Jadrana tijekom povijesti. Miscellanea Hadriatica et Mediterranea 1:169–197
Kozličić M, Faričić J (2016) Značenje Vira u pomorsko-geografskom sustavu Jadrana u prošlosti. In: Magaš D (ed) Otok Vir. Sveučilište u Zadru and Općina Vir, pp 255–276
Kozličić M, Faričić J (2004) The significance of Sveti Andrija Island (Svetac) on a sailing route across the Adriatic presented on old geographical maps. Geoadria 9(1):33–49. https://doi.org/10.15291/geoadria.128
Kozličić M, Faričić J, Uglešić S (2012) Geographical Basis of Navigation along the Velebit Channel according to the 1639 Senj Pilot. Geoadria 17(1):45–71. https://doi.org/10.15291/geoadria.237

Kozličić M, Mlinarić D, Andrić M (2015) The Zadar maritime zone in Pîrî Reis's Book of Navigation of 1526: maritime and geographical images of part of the Croatian Adriatic. Geoadria 20(2):119–147. https://doi.org/10.15291/geoadria.5

Kozličić M, Uglešić S (2015) The early 17th century Senj Pilot. The Velebit Podgorje region between Senj and Dračevac, according to Croatian and German editions of the Senj Pilot, University of Zadar

Kretschmer K (1909) Die italienischen Portolane des Mittelalters. Ein Beitrag zur Geschichte der Kartographie und Nautik. Veröffentlichen des Instituts für Meereskunde und des geographischen Instituts an der Universität Berlin

Krygier JB (1995) Cartography as an art and a science? Cartograph J 32(1):3–10. https://doi.org/10.1179/000870495787073762

Lago L (1998) Imago Adriae—La Patria del Friuli, L'Istria e la Dalmazia nella cartografia antica. La Mongolfiera Libri

Lago L (ed) (2002) Imago Italiae. Fabrica dell'Italia nella storia della cartografia tra Medioevo ed Età moderna. Realtà, immagine e immaginazione dai codici di Claudio Tolomeo all'Atlante di Giovanni Antonio Magini. Università di Trieste e Goliardica Editrice

Lago L, Rossit C (1984) Pietro Coppo. Le "Tabulae" (1524-1526). Una preziosa raccolta cartografica custodita a Pirano. Note e documenti per la storia della cartografia. Lint Editoriale Associati

Livieratos E, Boutoura C (2018) Carte Pisane and its coastline shape. e-Perimetron 13(3):161–181

Loomer SA (1987) A cartometric analysis of Portolan Charts: a search for methodology (doctoral dissertation). The University of Wisconsin

Marelić T (2020) Točnost prikaza Jadranskog mora na portulanskim kartama. Doctoral thesis. University of Zadar

Marelić T (2022a) Conformal cylindrical properties of Adriatic Sea Basin renderings on Portolan Charts. Cartograph J 59(2):83–101. https://doi.org/10.1080/00087041.2021.1956263

Marelić T (2022b) The origin of distance and bearing navigational data contained in portolani for the Adriatic Sea basin. Int J Cartograph (published online 01 April 2022):1–23. https://doi.org/10.1080/23729333.2022.2047402

Marelić T (2023a) Geometric stalemate and de-evolution of Adriatic Sea representations on early modern age nautical charts. Cartograph J 60(3):216–229. https://doi.org/10.1080/00087041.2023.2172533

Marelić T (2023b) Doubly perceived shape of the Adriatic Sea Basin on early modern geographical maps and nautical charts. Kartografija i geoinformacije 22(2):20–41. https://doi.org/10.32909/kg.22.39.2

Marelić T (2024a) Certain arguments against the hypothesis that portolan charts were genuine late-medieval cartographic products. Int J Cartograph 10(1):94–117. https://doi.org/10.1080/23729333.2022.2163146

Marelić T (2024b) Traces of the common origin of Carte Pisane, Cortona Chart and Pietro Vesconte's Charts. KN – J Cartograph Geograph Inf 74(2):137–156. https://doi.org/10.1007/s42489-023-00154-6
Marelić T (2024c) Copying-lineages of Portolan chart metrics and implications of their pre-Medieval origin. Int J Cartograph (published online 16 July 2024):1–26. https://doi.org/10.1080/23729333.2024.2349974
Marelić T (2025a) Mosaics within mosaics: the geometry of Portolan Atlases that contradicts the medieval origin hypothesis. KN – J Cartograph Geograph Inf 75(1):57–78. https://doi.org/10.1007/s42489-024-00176-8
Marelić T (2025b) Are Portolan charts and Portolan mile geometrically rooted in classical antiquity? A cartometric analysis of al-Shirazi's "Greek Map" and the Pisane, Lucca, Avignon, and Cortona Charts. Cartographica 59(4):143–160. https://doi.org/10.3138/cart-2024-0029
Marelić T, Faričić J (2023) Magnetic declination on the early modern cartographic representations of the Adriatic Sea. Abstracts Int Cartographic Assoc 6(155):1–2. https://doi.org/10.5194/ica-abs-6-155-2023
Marieni G (1830) Portolano del Mare Adriatico, Compilato sotto la Direzione dell'Istituto geografico militare dell'I. R. Stato maggiore generale. Imperiale regia stampera
Marković M (1993) Descriptio Croatiae—Hrvatske zemlje na geografskim kartama od najstarijih vremena do pojave prvih topografskih karata. Naprijed
Matković P (1862) Alte handschriftlische Schifferkarten in den Bibliotheken zu Venedig. Mitteilungen der Geographischen Gesellschaft in Wien:79–107
May WE (1973) A history of marine navigation. W. W. Norton & Company
Mirošević L, Faričić J (2015) Reflections of political-geographic shifts in the use of the geographic name 'Dalmatia' on maps in the early modern period. Annales—Series historia et sociologia 25(4):845–860. UDC 912.43(497.58):81'373.21
Mlinarić D (2012) A culture of cartography: maps of the Central Dalmatian Coastland. IMCoS J 129:15–20
Mlinarić D (2014) Stare karte Dalmacije promatrane kroz prizmu neokartografije: Čitanje povijesno-geografskih reprezentacija 16. i 17. stoljeća. In: Mirošević L, Graovac Matassi V (eds) Dalmacija u prostoru i vremenu: Što Dalmacija jest a što nije?. Sveučilište u Zadru, pp 89–100
Mlinarić D, Gregurović S (2011) Kartografska vizualizacija i slika Drugoga na primjeru višestruko graničnih prostora. Migracijske i etničke teme 27(3):345–373
Mlinarić D, Miletić Drder M (2017) Zbirka Novak: Mappae Croaticae u Zbirci zemljovida i atlasa NSK. National and University Library in Zagreb
Mlinarić D, Faričić J (2024) The Eastern Adriatic Coast in the narratives of the early-modern ottoman navigator and cartographer, Piri Reis. In: Ak M, Üstüner A (eds) Geography and Cartography in the Ottoman Empire. Istanbul University Press, pp 381–402. https://doi.org/10.26650/B/AH03AH08.2024.039.012

Monmonier M (1996) How to lie with maps, 2nd edn. University of Chicago Press
Monmonier M (2004) Rhumb lines and map wars: a social history of the Mercator projection. The University of Chicago Press
Nicolai R (2014) A critical review of the hypothesis of a medieval origin for Portolan Charts (doctoral dissertation). University of Utrecht
Nicolai R (2015) The premedieval origin of Portolan charts: new geodetic evidence. ISIS 106(3):517–543. https://doi.org/10.1086/683532
Nicolai R (2016) The Enigma of Portolan Charts—a geodetic analysis of the hypothesis of a medieval origin. Brill-Leiden
Nicolai R (2018) Analysing map analyst and its application to Portolan charts. e-Perimetron 13(3):121–140
Nicolai R (2024) The origin problem of nautical cartography: the importance of evidence and method. Int J Cartograph 1–25. https://doi.org/10.1080/23729333.2024.2352822
Nordenskiöld AE (1897) Periplus: an essay on the early history of charts and sailing-directions. P. A. Norstedt & Söner
Novak D, Lapaine M, Mlinarić D (eds) (2005) Five centuries of map and charts of Croatia. Školska knjiga
Panofsky E (1970 (1955)). Meaning in the visual arts. Penguin.
Presciuttini P (2004) Coste del Mediterraneo nella cartografia europea: 1500-1900. Priuli & Verlucca
Pujades i Bataller RJ (2007) Les cartes portolanes: la representació medieval d'una mar solcada. Lunwerg Editores
Rajaković M, Kljajić I, Lapaine M (2014) Map Projection Reconstruction of a Map by Mercator. In: Buchroithner M, Prechtel N, Burghardt D (eds) Cartography from pole to pole lecture notes in geoinformation and cartography. Springer, pp 31–44
Ribeiro DM, Caquard S (2018) Cartography and art. In Wilson JP (ed) The geographic information science & technology body of knowledge. https://doi.org/10.22224/gistbok/2018.1.4
Rossit C, Selva O, Umek D (2006) Imago Adriae—L'Adriatico e l'Abruzzo nelle antiche carte geografiche. Editrice Sigraf
Selva O (2013) Lo stato della cartografia veneziana tra XVI e XVIII secolo: emblema di potere e strumento di pianificazione territoriale. Bollettino dell'Associazione Italiana di Cartografia 148:69–87. http://hdl.handle.net/10077/12268
Sheehan KE (2012) Utility and aesthetic: the function and subjectivity of two fifteenth-century portolan charts. The Portolan 83:7–23
Shipley G (2011) Pseudo-Skylax's Periplous: the circumnavigation of the inhabited world: text, translation and commentary. Liverpool University Press Series. Bristol Phoenix Press

Slukan Altić M (2003) Povijesna kartografija: kartografski izvori u povijesnim znanostima. Meridijani

Slukan Altić M (1998) Pojava i razvoj portulana s posebnim osvrtom na portulanski atlas Diega Homena. Geodetski List 52(2):99–107

Slukan Altić M (2016) British contribution to the charting of the Adriatic Sea. Cartographic J 53(4):305–317. https://doi.org/10.1080/00087041.2015.1108659

Smyth WH (1854) The mediterranean: a memoir—physical historical and nautical. John W. Parker & Son

Snyder JP (1993) Flattening the earth: two thousand years of map projections. The University of Chicago Press, Chicago

Soja E (2001) Postmodern geographies. The reassertion of space in critical social theory. Verso

Stevenson EL (1911) Portolan charts—their origin and characteristics with a descriptive list of those belonging to The Hispanic Society of America, Publications of The Hispanic Society of America, 82. The Knickerbocker Press

Šakaja L (2015) Uvod u kulturnu geografiju. Leykam International

Taylor EGR (1957) The haven-finding art—a history of navigation from Odysseus to Captain Cook. Abelard-Schuman Ltd. and William Clowes & Sons

Tobler W (1966) Medieval distortions: the projections of ancient maps. Annu Assoc Am Cartographer 56:351–361. https://doi.org/10.1111/j.1467-8306.1966.tb00562.x

Tobler W (1977) Numerical approaches to map projections. In: Kretschmer E (ed) Studies in theoretical cartography. Deuticke, pp 51–64

Todorova M (2015) Imaginarni Balkan, Naklada Ljevak.

Tooley RV (1962) Maps and map-makers. Bonanza Books

Turnbull D (1996) Cartography and science in Early Modern Europe: mapping the construction of knowledge spaces. Imago Mundi 48:5–24. https://doi.org/10.1080/03085699608592830

Vagnon E (2007) Cartes marines et réseaux à la fin du Moyen Âge. In: Coulon D, Picard C, Valéria D (eds) Espaces et Réseaux en Méditerranée VIe—XVIe siècle, Vol. I La configuration des réseaux. Éditions Bouchène, pp 293–308

Vagnon E (2021) Representations. Painting, mapping, re-inventing the seas of the world. In: Lambourn E (ed) A cultural history of the sea II. Bloomsbury Publishing, pp 159–192

Valerio V (2006) Landscapes and charting in the nineteenth century Neapolitan-Austrian and English cooperation in the Adriatic Sea. In: van Gestel-van het Schip P, van der Krogt P (eds) Mappae Antiquae—Liber Amicorum Günter Schilder. HES & De Graaf Publishers, pp 465–478

Stouraiti A (2013) Talk, script and print: the making of island books in early modern Venice. Hist Res 86(232):207–229. https://doi.org/10.1111/1468-2281.12014

Tolias G (2007) Isolarii, fifteenth to seventeenth century. In: Woodward D (ed) The history of cartography, 3: Cartography in the European renaissance. University of Chicago Press, pp 263–284

Wagner H (1896) The origin of the mediaeval Italian nautical charts. Report of the sixth international geographical congress, pp 695–705. Royal Geographical Society, reprint in: Acta Cartographica, 5 (1969), 476–485

Wintle M (1999) Renaissance maps and the construction of the idea of Europe. J Hist Geograph 25(2):137–165. https://doi.org/10.1006/jhge.1999.0111

Woodward D (2007) Cartography and the Renaissance: Continuity and Change. In: Woodward D (ed) The history of cartography, Vol. 3, Part 1—Cartography in the European Renaissance. University of Chicago Press, pp 3–24

CHAPTER 2

Cultural Directions for the Adriatic Sea, or the Gulf of Venice

Patrizia Licini de Romagnoli

1 Introduction

Given the difficulty ships faced in trying to leave the Venetian lagoon, Napoleon needed another port to house his French fleet. Initially, he turned to the port of Ancona, having advocated to the Directory of France in 1797 that "France should retain this port as it would render us masters of the Adriatic Sea. Due to Ancona's open harbour, ships could enter and leave there under all winds" (Johnson 2016: 443). Napoleon entered Papal Ancona on February 10, 1797, and on February 19, he compelled Pope Pius VI to recognize the surrender of Avignon, both the Legations of Bologna and Ferrara, Romagna, and Ancona, all subject to Rome (Ciavarini 1867: II, 200–202).

There was a growing concern on the other shore of the Adriatic. Vessels no longer suited for use had been arriving for two days when Andrea Querini, *per la Serenissima Repubblica di Venezia Procuratore generale in Dalmazia e Albania*, sent the alarming dispatch, No. 55, of February 13, 1797, to the Most Serene Prince, Ludovico Manin, the hundred-and-twentieth and last Doge of Venice, informing him of the sudden

P. L. de Romagnoli (✉)
Associazione dei Geografi Italiani (A.Ge.I.), Milan, Italy

J. Faričić, T. Marelić (eds.), *Early Modern Nautical Charts of the Adriatic Sea*, https://doi.org/10.1007/978-981-96-8046-7_2

appearance of Pontifical vessels (*Bastimenti Pontifizj*) on the shores of Zadar. Now, not only was Querini sure that the French would invade the States of the Holy See, but also that this news spread consternation among all classes of people in Ancona at the ferocity of the aggressor forces. One hundred and twelve miserable fugitives have been landing on our shores lately, he writes to the Doge. They are all looking for a safe haven here in Zadar. Querini had received feedback indicating that many ships from Ancona, and from diverse cities of Romagna, had hastily weighed anchor and stood out to sea in different directions. Anconitan ships succeeded in reaching Istra (*Istria*), Lošinj (*Lossin*), and the lower ports of Dalmatia in safety. Setting forth that so many wretched people were, in and prior to the year 1797, engaged in regular and legitimate trade and commerce with the subjects of Venice, Querini has the fullest confidence that the amity previously existing between Ancona and Dalmatia would continue uninterrupted. Human rights would prevail. On March 2, 1797, the Signoria of Venice received the dispatch (VEAS, Senato, Dispacci: b. 467).

Realizing that he could not hope to take Vienna unaided, Napoleon decided on a bluff. He advanced as far as Leoben, just seventy-five miles from Vienna, and then offered a truce. The Austrians agreed a five-day cessation of hostilities. Preliminaries of peace were signed at Leoben on April 18. As a sweetener to get Imperial Austria to accept the Leoben terms, Napoleon had included a secret clause promising that the Habsburg Holy Roman Empire could swallow up the neutral territory of the *Serenissima*. Napoleon was never more unscrupulous than in his treatment of Venice in 1797. On the other hand, a direct attack on Venice might drag his Army in Italy into a prolonged siege, since the *Veneto Dominio* could easily be reinforced and provisioned by sea.

Meanwhile Venice's harbour was closed to foreign ships, under armed neutrality. For 1342 years Venice had been able to preserve independence from the Reign of Italy. On April 20, a French ship—the *Libérateur de l'Italie*, no less—sailed into the Venetian lagoon, where its captain Jean-Baptiste Laugier apparently pressed on. Five Frenchmen were killed, including Laugier. And then, on May 1, Napoleon indignantly sent an ultimatum to the Doge and the Grand Council of Venice calling for immediate surrender at the threat of destroying Venice. Notwithstanding the neutrality of the said port, as commander-in-chief of the French fleet in Italy, a furious Napoleon issued, at Palmanova, a violent manifesto, in which he accused the Senate of Venice of having perfidiously encouraged the people to murder the French in the dastardliest manner and declared that a government guilty of so many outrages ought to be treated as the determined enemy of France.

Napoleon was right in placing the issue of the ten Regiments of Schiavoni (*les Esclavons*) first in the manifesto. He then commanded his generals of divisions to treat those of Venice as enemies, and to pull down, in every town of the *Stato da Terra*, the Lion of St Mark (Tentori 1799: II, 352).

Palmanova is the star stronghold at the mouth of the Adriatic Sea along the administrative inner border between the *Stato da Terra* and the *Stato da Mar* of the *Serenissima* with the eastern battlements towards the outer border with Habsburg Austria. The Senate chose the date of October 7, 1593, as the founding date, because it marked the victory of the Christian forces over the Ottoman Turks in the Battle of Lepanto.

By June 15, the Grand Council of Venice was called for the ratification of these degrading terms. Nicolò IV Morosini, the last Duke of the Venetians, or, as they called him, the *Doge*, exhorted the nobles to resign their power. The Grand Council thus decided to return sovereignty to the people, in favour of a provisional Municipality. After 1342 years the Republic of Saint Mark, which had fought off Spaniards and Turks, renounced its independence without offering any further resistance.

Since July 20, 1796, the Grand Council of Zadar had invoked a rapid occupation of Dalmatia on the part of Imperial Austria to put an end to domestic rebellions in *Traù* (Trogir), *Sebenico* (Šibenik), and *Lesina* (Hvar) according to the letter of Lelio Cippico, Metropolitan Archbishop of *Spalato* (Split) (Kolumbić 1979: 104; Agostini 1998: 582–583). Meanwhile, and on June 15, 1797, the so-called Four Days of *Cherso* (Cres Island, modern Croatia), crowned by Venetian Lions, ended with Habsburg field marshall Count von Klenau quitting the rebellion. On June 19, in *Pola* (Pula, modern Croatia), an ordinance was issued permitting the Provisional Court of Justice (*Superiorità Provisoria*) formed by Domenico Licini and Antonio Tromba, judges, Domenico dalla Zonca and Cristoforo Tromba, deputies, and Giorgio Davanzo and Andrea Trevisan, the mayors of *Dignano* (Vodnjan, modern Croatia), to enact extraordinary measures concerning the administration of the city of *Dignano*, until the Cesarean Royal Commissary (*Cesareo Regio Commisario*), Raymond the count von Thurn, came up (Fig. 2.1). The document, which is here published for the first time, was signed in Pula by the Cesarean Royal Major-General and commander of the Vodnjan garrison, Count von Klenau. Domenico Licini and Cristoforo Tromba had

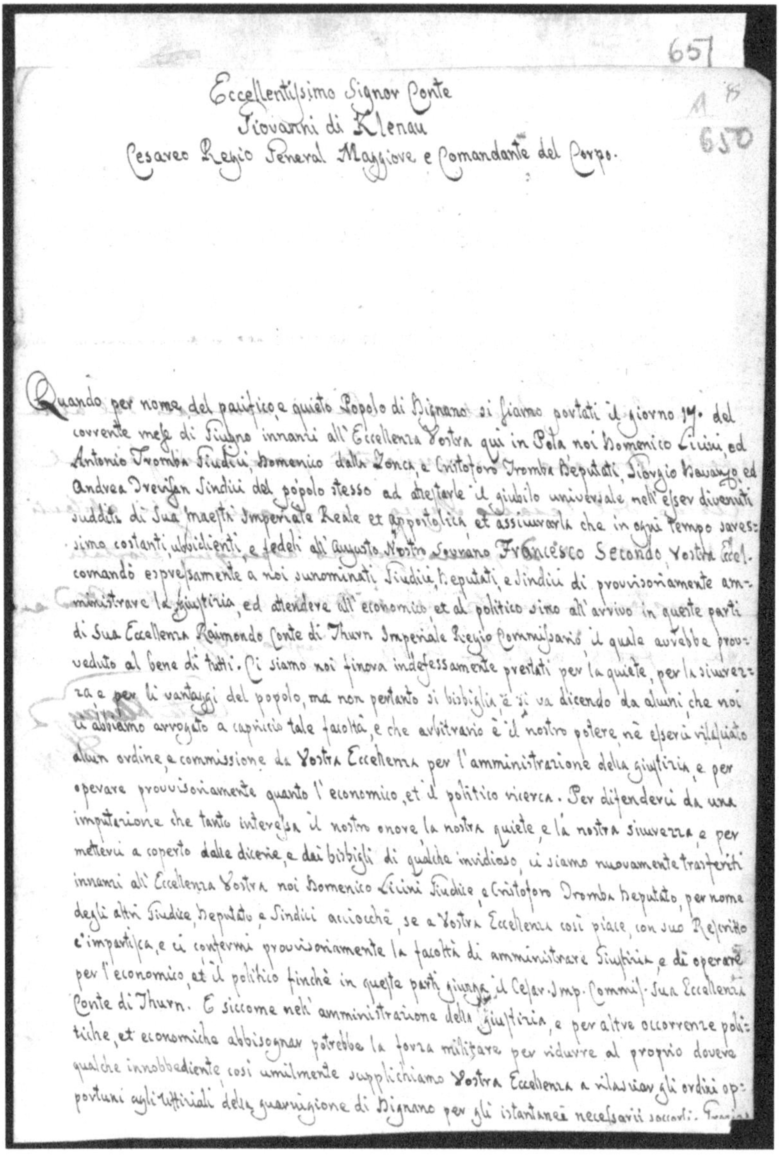

Eccellentissimo Signor Conte
Giovanni di Klenau
Cesareo Regio General Maggiore e Comandante del Corpo.

Quando per nome del pacifico e quieto Popolo di Dignano si siamo portati il giorno 17. del corrente mese di Giugno innanzi all'Eccellenza Vostra qui in Pola noi Domenico Licini, ed Antonio Tromba Giudici, Domenico dalla Zonca, e Cristoforo Tromba Deputati, Giorgio Bavanzo ed Andrea Drevisan Sindici del popolo stesso ad attestarle il giubilo universale nell'esser divenuti sudditi di Sua Maestà Imperiale Reale et Appostolica et assicurarla che in ogni tempo saressimo costanti, ubbidienti, e fedeli all'Augusto Nostro Sovrano Francesco Secondo, Vostra Eccel. comandò espressamente a noi sunominati Giudici, Deputati, e Sindici di provvisoriamente amministrare la Giustizia, ed attendere all'economico et al politico sino all'arrivo in queste parti di Sua Eccellenza Raimondo Conte di Thurn Imperiale Regio Commissario il quale avrebbe provveduto al bene di tutti. Ci siamo noi finora indefessamente prestati per la quiete, per la sicurezza e per li vantaggi del popolo, ma non pertanto si bisbiglia, e si va dicendo da alcuni che noi ci abbiamo arrogato a capriccio tale facoltà, e che arbitrario è il nostro potere, nè essersi rilasciato alcun ordine, o commissione da Vostra Eccellenza per l'amministrazione della giustizia, e per operare provvisoriamente quanto l'economico, et il politico ricerca. Per difenderci da una imputazione che tanto interessa il nostro onore la nostra quiete, e la nostra sicurezza, e per metterci a coperto dalle dicerie, e dai bisbigli di qualche invidioso, ci siamo nuovamente trasferiti innanzi all'Eccellenza Vostra noi Domenico Licini Giudice, e Cristoforo Dromba Deputato, per nome degli altri Giudice, Deputato, e Sindici acciocchè, se a Vostra Eccellenza così piace, con suo Rescritto c'impartisca, e ci confermi provvisoriamente la facoltà di amministrare Giustizia, e di operare per l'economico, et il politico finchè in queste parti giunga il Cesar. Imp. Commiss. Sua Eccellenza Conte di Thurn. E siccome nell'amministrazione della Giustizia, e per altre occorrenze politiche, et economiche abbisognar potrebbe la forza militare per ridurre al proprio dovere qualche innobbediente così umilmente supplichiamo Vostra Eccellenza a rilasciar gli ordini opportuni agli Ufficiali della guarnigione di Dignano per gli istantanei necessarii soccorsi. [illegible]

Fig. 2.1 Ordinance of the Cesarean Royal Major-General in Pula and commander of the Vodnjan garrison, Count von Klenau, dated June 19, 1797, for Vodnjan (Source: Trieste, Archivio di Stato, Atti amministrativi dell'Istria 1797–1813, busta 47, c. 650–651). Courtesy of the Library (09.09.2015) come in to share in the general joy and jubilation of the people of Vodnjan in becoming subjects of his Imperial, Royal, and Apostolic Majesty, Francis II of Austria (TSAS 1797–1813: busta 47, c. 650–651).

Then came Count Lorenzo, Licini in Zadar and Rubcich in Poljica, Domenico's father, official surveyor of Dalmazia since 1762 (ZDAZ, 1781–1783: I; 1783–1785: I; 1763–1765, I, f. 241 a-b). The Grand Council of Zadar had deputed him to give minute description of the ceremony of the deposition of the Venetian flags at the altar in the Cathedral church at Zadar on July 1, 1797. The presiding officer was Sergeant Major General Antonio Stratico, surrounded by Venice's Militia. Fragments of Lorenzo's account entitled *Oh giornata fatale, critica e dolorosa!* make it clear that it was not an easy day for Dalmatia (Dandolo 1855: 266–267). The Rubcich family was named Licini in the city of Zadar (Pavich 1907: 311). The family took the name Rubcich from the place in Poljica, historical Croatia, where Rubich was a feudal residence.

On October 17, 1797, Napoleon rapidly signed the peace at Campo Formio with Austria.

This chapter addresses the Adriatic names on two essentially unstudied nautical charts by Pareto (active 1453–1455) and Angelo de Conte Freducci (active 1547–1556). These place-names are substantial departures from traditional nautical chart place-names. Pareto's 1455 chart is in Rome (Biblioteca Nazionale Centrale, Carte nautiche 1). Angelo Freducci's 1556 atlas is in Mantua (Biblioteca Comunale Teresiana, MS 646).

As we shall see, we can use the said place-names as a key to understanding history. An examination of the Adriatic names and their stories will shed light also on the reason why Napoleon decided in 1797 that a direct attack on Venice might drag his Army of Italy into a prolonged siege. In exile at Saint Helena, Napoleon, recollecting the affairs of Venice, went into detail in his conversations with Las Cases (1823: vol. 2, t. IV, p. 31). Emmanuel de Las Cases was with Napoleon during his captivity, between October 1815 and November 1816, and was the author of the celebrated narrative, *Mémorial de Sainte-Hélène*. Finally, as our conclusion to this analysis, we shall use Napoleon's words in his conversations in the *Mémorial*.

2 Travelling Back in Time Is Possible

To the historical geographer of late medieval and early modern European cartography the portolan charts and their routes plotted graphically to form networks are fundamental documents. Chartmakers during the fifteenth and sixteenth centuries added new links to the traditional sites on charts to be used for navigation, according to continuous civilization's evolution along a cultural line. Nautical charts arrayed sequentially thus provide a short cut through cultural heritage in a well-identified global navigational system.

Turning to the content of both single charts and atlases, the most obvious of the common denominators that link the earliest survivor (the Carte Pisane), probably a little before 1300, to those of several centuries later is the network of interconnecting rhumb lines (Campbell 1987: 371, 372, 376, 426). Both the geographical area they cover and most of the graphical elements (rhumb lines, scale, images of sovereigns, etc.) are traditional. However, beneath the superficial similarity there are fundamental differences in legends and toponyms, perhaps establishing that one might have been drawn with information gained differently than the other. Contrary to expectation, the toponymy of the Mediterranean was under constant review by the portolan chartmakers.

In the northeast is an arm of the Mediterranean, called the *Adriatic Sea*, or the *Gulf of Venice*. It is about 600 miles long, and 150 wide, and its narrow entrance is commanded by the island of Corfu, the key to all the countries under the administration of the now-vanished *Serenissima*. It has several excellent harbours, but in some parts the coast is dangerous. And, indeed, sailing directions for the Adriatic Sea are not sufficient to guide a ship pilot without the assistance of navigation charts with rhumb line courses drawn on them. By the late thirteenth century navigation charts were being produced to accompany the handbooks. The archival aspect of navigation charts continues the Roman and medieval coordinate systems for surveying and recording plots of land (Casey 2002: 177). As T. Campbell further points out, from the earliest extant copies, probably a little before 1300, the outline they gave for the Mediterranean was amazingly accurate. In addition, their wealth of place-names constitutes a major historical source.

Portolan charts also combine archival documentation with sheer utility. A close scrutiny of the place-names on the chart made by Bartolomeo Pareto in Genoa reveals detailed geographical and political data not available elsewhere (Fig. 2.2). The signature "Presbiter Bartolomeus de Pareto Civis Januae Accolitus sanctissimi / domini nostri papae composuit hanc cartam MCCCCLV i[n] Janua" is clearly visible on Pareto's 1455 chart. A couple of years had passed since Constantinople and the Roman Empire of the East fell before Mehmed II. The defence of the *Stato da Mar* was then undertaken by reinforcing the defences of the system of naval bases in Venetian Dalmatia between 1467 and 1475 (e.g. VEAS, Senato-*Mare*: reg. 15, fol. 14). It should be remarked that the designation *Magnus turcus* refers specifically to Mehmed II, who was the first of the Ottoman emperors on whom the Christians bestowed the title of The Great Turk; it was not owing to his great actions that this splendid title was accorded to him, but to the vast extent of his dominions, in comparison with those of the Sultan of Konya (Lat. *Iconium*) in the central Anatolian Plateau, his contemporary, who was distinguished by the title of The Little Turk (D'Israeli 1793: 348–349). The earliest chart to use this title and illustration is Pareto's chart of 1455—quite soon after the event.

The chart was made by Pareto for Pope Nicholas V. Both came from the State of Genoa in the Reign of Italy. The Pope's coat of arms with the two initial letters of his name (i.e. *NV*) is emblazoned in its proper colours. It is undoubtedly true that since 1453 Pareto, a Genoese, acolyte of the Pope, had taken advantage of his official position at "The House of public debts and banks (plural) of St George," usually translated as The Bank of

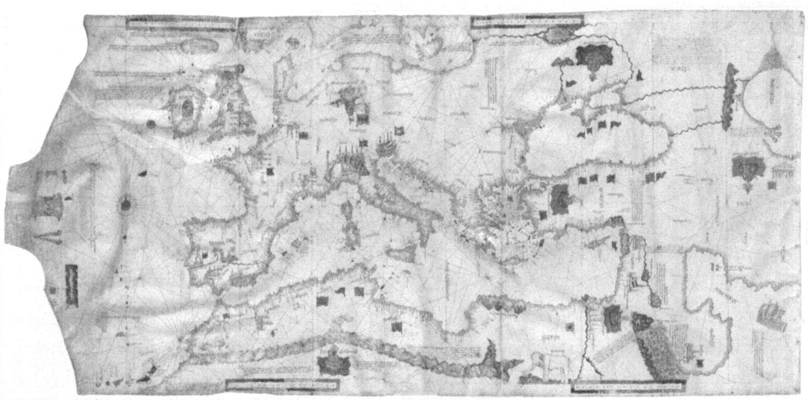

Fig. 2.2 Bartolomeo Pareto's 1455 luxury nautical chart (Source: Rome, Biblioteca Nazionale Centrale, Carte nautiche 1) © BNC Board

St George at Genoa, the greatest and most powerful holding company in the entire world (Belgrano 1866: 494).

The sequence of the settlement names of the eastern Adriatic coast in 1455 runs thus: Venice (*Venecıa*), Croatia (*Crovacıa*), Albania (*Albanıa*), Skanderbeg (*Scandrebecco*). Skanderbeg, or the Great Alexander (Iskander-beg), was named to represent the State of Scander-begh, the settlement of an exhausting war between *Georgius Castriotus nominatus Scanderbecus* (1405–1468), the Albanian *condottiere*, and the invading Turks. As far as we know, the land of George Castriota nicknamed Scander-begh appears on charts for the first and last time. A crushing defeat at the Battle of Svetigrad (modern Kodžadžik, North Macedonia) in Scander-begh's dominion towards Durazzo led him to ask Pope Nicholas for money in 1449. The Archbishop of Antìbari (modern Montenegro), however, gave a vague answer in 1450. Dubrovnik (Ragusa) traders found the money needed to raise an army. Ten years had passed since the defeat of the Christian coalition at Varna on the Black Sea in November 1444 when men who had attacked Belgradians and Italians at the camp of Belgrade on June 25, 1454, were now devastating the country and meeting with no resistance until Scander-begh got a few Albanians together to resist any further advance. The fugitives carried the alarm to Rome. But the Oltremarina war, as chronicles called it, again began with vigour. This was followed by an attempt to assassinate Scander-begh, and this again by a conspiracy among his nobles headed by Moisè of Dibra at the border with North Macedonia (modern Dibër, Albania) (Biemmi 1756: Lib. III, 154, 204–227; Lib. V, 319–347). For several years the Albanian feudal lord had been fighting alone in Albania against the deadly foes and now he needed the support of the Pope more than ever when Pareto made his chart in Genoa in 1455.

Further, *Crovacıa*, a variant spelling of Croatia, may be philologically related to the word *krv* which means "blood" in the Croatian language, as it was supposed to be the genuine native place-name. Native peoples use place-names and their stories both as vital information and a key to their identity. But the place-name was replaced sometime during the mid-fourteenth century by *Sclavonıa*, a corruption of *Slavonıa*. The great number of villages scattered across the Slav countryside are known by the name of *Sclavonȷa* (dotless j) on Dulcert's charts of 1339. The error of "sclavus" (slave) for "slava" (glory), however, goes back to the Carolingian Empire. It was not until the ninth century that in Germany a new Latin word for "slave" appeared alongside the classical *servus*, *ancillus*, and *mancipium*:

sclavus/sclavi (Reuter and Nelson 2006: 249). According to Lučić, Slavs occupying Dalmatia were Croats and Serbs; the altered name of "Sclavos" for "Slavs" as a general appellative was then a foreign adaptation of *Serbus* and *Sclavus*, from *servus*, the new Latin word, to indicate both Serbs in *Servia* (modern Serbia) and Croats in *Sclavonia* where the lower part was called Dalmatia (Lučić 1666: Lib. VI, Cap. IV, De Sclavonia: 278–281).[1]

3 The Armed Fleet of St Cyriacus at Anchor in the Port of the City of Ancona in the Papal States

If there has been disagreement about the function of the separate charts, it has been generally supposed that volumes of nautical charts were intended from the outset for the library shelf (Almagià 1944: l, VIII and note 35). The little that can be learned of Anconitan chartmaking tradition, however, advises caution in this respect. Fortunately, some unknown manuscripts have been found in Loreto, documenting the activity of Angelo Freducci (active 1547–1556) as a *regolator* of the City of Ancona in 1550–1551.[2] Moreover, on March 9, 1563, Giovanni Freducci, the son of Angelo, was declared a Knight of the Order of St Stephen, vested at Pisa. Letters Patent under the great seal of the City of Ancona in the Reign of the Pope, bearing dates February 11 and 12, 1563, had been necessary to prove his nobility; and consequently a pedigree of three generations of the Freducci family was placed on Official Record at the Military Order of St Stephen in Pisa (LOASSC, 1550 and 1551: Ancona, cc. 32vD, 33rD; Licini 2020: 34–35, 104–105, 142).

The quantity of this information is sufficient to allow a sufficiently precise conclusion to be drawn. Three atlases, 1554 (Warsaw), 1555 (Greenwich), and 1556 (Mantua), have been compiled consecutively. Under the Statute of the City of Ancona (printed 1566), on the first of April of each year as a collector agent of the Commune (*Regulatori Communis Anconae*) Angelo Freducci was the officer who had the power to examine ship's captains. It was Angelo's duty to appoint a captain of the armed fleet of St Cyriacus belonging to the Commune of Ancona, within the Ecclesiastical States.

[1] See also Orbini, 1601: *DALMATIA, CROATIA, BOSNA, SERVIA, RASSIA, ET BULGARIA*. Mavro Orbini is one of the most prominent Dubrovnik authors.

[2] *Libro di spese diverse* is a transcription of Lorenzo Lotto's book of accounts, 1538–1556, with commentary, chronology, glossary, and indexes.

The role of *Regulatori* does not stop there, though. The Anconitan nobles are divided into classes. The *Regulatori* occupy the highest rank (Ferretti IB 1556: Collatio Prima, De Officiis: Rub. 8, *De Officio Syndicatorum*, 7–8; Rub. 35, *Ordo pro Armata Sancti Chiriaci*, 26–27; Rub. 110, *Quod non cassetur*, 76; Ferretti 1685: 152, *Regolatori della Città, Supremo Grado de Nobili*).

Since the Norman era (June 1080), Rome had been the capital of the Ecclesiastical States, until they were incorporated into a unified Reign of Italy in 1870 (Caspar 1923: 515–516; Weiland 1893: 554–556). Both linguistic and geographical features appear relevant factors worthy of consideration when thinking about cross-Adriatic routes. Jakinska Bay in the island of Molat took its name from Jakin, the old word in the Croatian language for Papal Ancona, and the church and harbour of St Cyriacus in Premuda, the outmost island in the Zadar Channel, were named after the patron saint of Ancona (Faričić 2014: 36). Premuda and Molat are the two stations situated at the coterminal points on the actual line of route from Ancona to Zadar. Since 1379 the King of Hungary had planned an emergency rescue route across the Adriatic from Papal Ancona to the stronghold of Zadar (Ferretti 1685: 55).

The atlas by Angelo Freducci, namely Mantua, Biblioteca Comunale, MS 636, dated 1556, is stylistically very similar to the other two atlases. Map 6 of Angelo's atlas of 1556 covers the region around the Adriatic. The signature reads "Angelus eufredutius Anconitan[us] composuit Anconae In anno d[omi]ni MDLVI" (Fig. 2.3). The general system, on this scale, has been extended by the completion of two maps of the west coast of America, covering, respectively, the coast between the Antilles Islands of the Caribbean Sea down to Tobago and Trinidad (*Tobago-Trinidade*) and continental Venezuela (Map 1), and the coast of Brazil stretching between Cape St Agostino (*C. de Scõ Agustino*) and the harbour of Bahia de Todos os Santos (baia de totosdos = baia de totos sa[nto]s) (Map 2).

Map 1 is a luxury nautical chart, and the mountains are finely touched in gold. All the Balkan range is called *DALMATIA*. Painted images of mountains were superfluous for charts to be used for navigation, but at least in this case they were no extra-cost elements reserved for luxury charts to be owned and displayed by royalty or nobles. Even if Angelo, the son of Count Giovanni, was born in Ancona, he was a Freducci. In the middle and early modern ages, the *Comitatus Firmanus* was under the dominion of the Freduccis, a powerful family of maritime counts in Fermo

Fig. 2.3 Map of the region around the Adriatic, from an atlas by Angelo Freducci (Source: Mantua, Biblioteca Comunale Teresiana, MS 646, Map 6; atlas dated 1556). Courtesy of the Library (18/2019)

near Jesi, both of which towns formed part of the Marca of Ancona. It was about the year 1450 that through marriage with the powerful family of the Ferrettis, counts of Castel Ferretto and Torrette in the *Comitatus Anconitanus* (Ferretti 1685: 132–133), Ludovico the Elder, Freducci, Count of Falerone in Fermo Counties and senator in Rome, the first chartmaker in the Freducci chartmaking dynasty, became possessed of the Ancona citizenship and was admitted into the order of the Anconitan nobility (Licini 2020). Freducci and Ferretti were feudal lords of the Apostolic See respectively.

The said Letters Patent describe Giovanni Freducci as a man of the noble rank in Ancona and declare that the Freduccis never engaged in the vile mechanical arts. The descents are annexed to the Letters Patent; no one can be admitted to the Knights' list if they are unable to demonstrate four descents and display a valid coat of arms in accordance with that. The proofs of nobility for four descents on both sides (Freducci, Toroglioni, Orvieti, Scalamonti) must be made by arms and registers, signed by the notary-public within whose jurisdiction the candidate was born, by whom they are sent to the Grand Master of the Military Order of St Stephen, Cosimo I Medici, and there undergo another examination by the Knights of that Military Order in Pisa, who have the power to reject them.

It turns out that the purpose of Anconitan charts and atlases signed by a chartmaker as a member of the Anconitan patriciate was not that of merchandise. As a *regolator* of the Commune of Ancona (always a nobleman of the first rank), Angelo did not produce nautical charts available for public sale and refrained from any commercial activity. Sixteenth-century Ancona and its subject cities saw nobility as a caste defined by non-participation in vile mechanical arts. Maritime counts were commanded to look to the custody of the sea-coasts, as expressed in Charlemagne's *Capitularia*, Book IV, Chap. V, *De Comitibus ad custodiam maritimam deputatis* (Licini 2020: 40–47, 144).

As a professed Knight of the Order (1563), and Commendator of the Order residing in Ancona (1568–1569), Giovanni Freducci's chief duty was to participate in the Order's maritime campaigns to rescue Christians who had been captured by Infidel forces (Medici 1562: 2, Regola dei Cavalieri di Santo Stefano, Tilolo Primo, III, Dell'Uffitio, et professione de' Cavalieri). Giovanni's vision was also heraldic: he was ordered to procure the coats of arms of the Knights in painted wood. The Knights assembled in front of their escutcheons; the places of those who had died remained empty. Each escutcheon belonged to a different family and was correspondingly different. But all conformed to a standardized format and featured the red cross of St Stephen. Captain Francesco Ferretti, an Anconitan, Knight of the Order of St Stephen (since July 2, 1562), Freducci's cousin (four descents: Ferretti, Scalamonti, Bonarelli, Todini), added an interesting statement about Anconitan chartmaking in 1579. Back from the victorious Battle of Lepanto against the Ottoman Turks in 1571, Ferretti composed *Diporti Notturni. Dialoghi Familiari*, a series of dialogues, adapted to an *isolario*, published in 1580 (engraved Tables, by Michel'Angelo Marrelli, an Anconitan, 1579). In the dialogue on Chorograpy, Geography, and Cosmograpy, Ferretti proudly maitains that the Freduccis, Anconitan gentlemen, his compatriots, made such good charts that, even in his days, they feared no rivalry from the most skilful noble authors of Genoa and Portugal. Now, he adds, Bartolomeo Bonomini (*Buon'huomini*), a compatriot, a most virtuous maker of charts in Ancona, is excellent in the art. Ferretti reminds that he presented a chart by Bonomini to Guidubaldo II della Rovere fifth Duke of Urbino and Lord of Pesaro. At the time of Pope Paul IV (1555–1559), Count Ferretti was a Company Sergeant Major of the Duke of Urbino's Regiment; he was put under the orders of Aurelio Fregoso, during the campaign of Naples in 1556 (Ferretti 1580: Dial. II, 9–11; Dial. V, 48; Ferretti 1685: 499–501; 322–324).

It will be useful to add that Aurelio Fregoso, a Genoese, whose grandmother Gentile was the sister of Guidubaldo II, Duke of Urbino and Lord of Pesaro, lived honoured in Florence with Cosimo Medici, 1st Grand Duke of Tuscany, 1st Grand Master of Knights of the Order of St Stephen. Francesco da Montauto, nicknamed Otto, son of Count Alberto Barbolani da Montauto (Arezzo), was Fregoso's son-in-law. Pirro, son of Francesco, received the Order on May 31, 1562 (PPC: 285; Dall'Ara 2020: 89, 122, 141–152, 204).

The subscription on Bonòmini's 1570 chart reads "Bartolomeus de Bonis Hominib.$^{\partial}$: Anc.^{s} Faciebat Ancon. 1570." ("Bartolomeus de Bonis Hominibus Anconitanus Faciebat Anconae 1570."), m. 0,93x0,54. He is also known by the name of Bonomi (Guglielmotti 1886: 36 n. 42). The detail of the Adriatic section displays the names *ISTRIA* and *CORVATIA*. The manuscript chart, however, is decorated also with images of cities and significant features of the hinterlands. Lake Scutari, "Lago de Scutari" on the chart, is richly detailed in the hinterland of the eastern Adriatic coastline.

There is some plausibility in attributing to Duke Marcantonio Colonna the property of the Anconitan chart by Bonomini, and now in Rome, Galleria Colonna (Fig. 2.4). The chart is not published, so far. It is true that Colonna, Duke of Pagliano, General of the Army of the Holy Church, one of the three commanders at the Battle of Lepanto, elected Francesco Ferretti of Ancona, son of Piergentile, as captain of one of the twelve galleys in Rome on June 12, 1570. Colonna at once got leave from the Pope and was able to reach Ancona via Castelnuovo di Porto, Terni, Serravalle, Macerata, and Loreto on June 19, 1570. Ferretti was stationed in Recanati, where he not only equipped his galley most beautifully and most perfectly, but also recruited men for the infantry regiments in the Army. Meanwhile Colonna received orders from the Pope, to leave Ancona, and sail for Venice in search of financial aid at the end of June; but he could not set sail from Ancona, on account of bad weather. Nevertheless, he took fifty

Fig. 2.4 Bonomini's 1570 chart (Source: Roma, Galleria Colonna, Tabula Nautica). Courtesy of the Galleria Colonna, Rome

cavalry men with him as guards and rode to Pesaro on July 1, 1570. Count Ferretti's role as Duke Guidubaldo's companion, undoubtedly, was to accompany Colonna, commander-in-chief of the combined naval forces, to the meeting with Guidubaldo II della Rovere, fifth Duke of Urbino and Lord of Pesaro. Therefore, a possibility is that Ferretti gave Bonomini's 1570 chart to Duke Guidubaldo during the mission from Ancona to Venice in July. The meeting is documented in private letters (Guglielmotti 1862: Lib. I, Cap. III, 14–22; Cap. IV, 22–25 and notes 16–17; Lib. II, Cap. IX, 170). It took place in the ducal residence named *L'Imperiale* outside of Pesaro on July 1, 1570. The company then left for Cesenatico-Fornaci and arrived at Chioggia and the villages of the Venetian lagoon. Old as he was, the Duke of Urbino obviously thought he could hand Bonomini's 1570 chart over to Colonna and Francesco Maria II della Rovere, his son and heir, as a trusty map to lead the way across the Adriatic Sea. Corfu, Cefalonia, and Lepanto, a small port of west Greece on the channel connecting the Gulfs of Lepanto and Patras, scene of the naval battle in 1571, are all mentioned in Bonomini's chart.

In August 1570 the twelve vessels were armed, eight of them at Papal Ancona, and four at Venice. Finally, in 1571 Count Francesco Ferretti, the head of the family of Ancona at the time, Knight of the Order of St Stephen, was duly attested for service in the Venetian Republic (Ferretti 1685: 500).[3]

The story had begun on the Tyrrhenian coast. Cosimo I Medici, Duke of Florence, went to Rome in February 1568, to receive the Grand-Ducal Crown bestowed on him by Pope Pius V. During that visit he is said to have pointed out to the Pope that the only way in which Christendom could make headway against the Turk was by a League between the Pope, Venice, and the King of Spain and the Two Sicilies. The Roman Church,

[3] Francesco Ferretti passed away on December 2, 1593, in Ancona. Guglielmotti makes a mistake when he says that the captain's name in 1570 was Alessandro Ferretti (1862: Lib. 1, Cap. III, 16–19; June 11, 1570, note 8). There is no record to show that Alessandro was in the line of descent from the Ferretti family of Ancona (Ferretti 1685). Corradino Astengo seems not to have observed the original chart by Bonomini (Astengo 2007: 180 and note 37, 222). Contrary to Astengo's argument, this chapter's conclusions are based on close observation of the original chart (now in the Galleria Colonna). The neck of the parchment excludes the possibility of its being part of a complete nautical atlas. Moreover, the date is clearly legible, and totally tenable in regard to private letters that were written by the protagonists themselves. Finally, Bonomini's chart does not come "to an end in the very area where the battle was fought," i.e. the Sea of Greece. It covers the whole Mediterranean Sea including the eastern area.

having no fleet, resolved to find another way, as suggested by Medici (Guglielmotti 1871: I, Libro I, Cap. IV, 23–5). When the League's forces began to be raised in 1569, Cosimo Medici, 1st Grand Duke of Tuscany, 1st Grand Master of Knights of the Order of St Stephen, between Pisa (Porto Pisano) and Leghorn (*Livorno*), caused to be fitted out twelve galleys, a royal galleon, a small *galeoncello*, and a frigate, and equipped them with everything needed for naval warfare. Of these vessels Pope Pius V paid only six galleys, the rest being sent to the aid of the League in Venice and Ancona at the expense of the Grand Duke himself (Baldini 1578: 65, 77–78).

The defeat of the Turks in the Battle of Lepanto (1571) then destroyed completely their ascendancy in the Mediterranean, according to Lord Francis Bacon in 1612 (*Of Kingdoms and Estates*).

Three coats of arms are found on the Mercator-Hondius map entitled SCLAVONIA, CROATIA, BOSNIA CUM DALMATIAE PARTE / GOLFO DE VENETIA PER GERARDUM MERCATOREM CUM PRIVILEGIO, fol. 303 (Kozličić 1995: 172–174; Novak et al. 2005; Rajaković et al. 2014). The volume on Italy and the Balkans was released in 1589 (Duisburg). Mercator, in describing the geography of Slavonia, Croatia, and Bosnia with a part of Dalmatia, accurately depicts three coats of arms representing, first, the Republic of Venice (the Lion of St Mark, the ducal crown of the Dogeship of Venice, or the horned cap); second, the Holy Roman Empire House of Habsburg ruling over Austria, Spain, Sicily, and the Reign of Italy (the double-headed black eagle, the Habsburg coat of arms, the imperial crown of the Roman Holy Roman Emperor); and third, the Ottoman Empire (the white star and crescent moon). The work is accepted as showing the geography of the so-called War of the Three Empires 1570–1571.

The rich connections between the Order of St Stephen and Ancona's cartographic legacy certainly deserve further study. The Knights of the Order of St Stephen were divided into two branches and had a house at Florence and a house at Ancona in the Papal States, on the Adriatic Sea (Ferretti 1685: 1562–1573, 322, 324; 1580–1587, 326–327). There is a final and important point to be made regarding the connections between the Order of St Stephen and the Adriatic Sea, or, as Volčić calls it, *Golfo di Venezia*, in his 1593 atlas in Naples (Helsinki, National Library): after the demise of the Anconitan chartmakers, which took place in the person of Bartolomeo Bonomini after his disappearance in the 1580s, one of the earlier charts by Volčić was dated Leghorn (*Liburni*), August 26, 1592. Volčic's chart of 1598 bears the coat of arms of the Barbolani da Montauto family, and painted over with the red cross of St Stephen. Vicko Dimitrije Volčić (1563–1607), the Dubrovnik chartmaker, one of the most known among Croatian authors, crossed the Adriatic in 1592 (Altić 2016, 2023). Did he pass through Papal Ancona?

In geographic terms, historical Croatia is bounded on the east by the Cetina (*Iilluro*) River Basin; and on the eastern bank of the Neretva (*Orontium/Narenta*) River begins the contiguous principality of Ragusa (*Ragusio*), modern Dubrovnik, where the people of that place are called Zachlumi. The principality extends north to Serbia, according to Lučić. A unified higher defence structure similar to that of the Anconitan Marca, which was then subdivided into the "City of Ancona and Comitatus Anconitanus" and the "City of Fermo and Comitatus Firmanus," seems to emerge in chapter 13 "De Zupaniis, sive Comitatibus Croatiae" upon the opposite shore of the Adriatic in the work of Lučić. As local self-government units, the eleven counties of maritime Croatia, Dalmatia, and what remained of Bosnia were called *županije* (Lučić 1666: XII, 47, XIII, 47).

The locution "Comitatibus Croatiae," however, was then less extensive in its application than at present: it designated, properly speaking, the area called *Kaštela Riviera* between the ports of Trogir (*Traù*) and Split (*Spalato*), or, as Venice called it, *Riviera dei Castelli* including the area of the Cetina River (*antiquitus Lilluro*) shaping Poljica territory, Primorje and Makarska, Klis (*Clissa*), and Imotski (*Imoschi*) (Foscarini 1722: 16–18; Licini 2009: II, 247–272).

The cultural system within which they were contributing to the establishment, extension, and maintenance of the navigable Adriatic Sea as a whole in the age of sail had existed since time immemorial. On the opposite shore of the Adriatic are Zadar, Šibenik, Split, and Dubrovnik. On this shore stand Pesaro, Loreto, Ancona, Ortona, which has the distinction of housing the skull of St Thomas, the twelfth and doubting apostle, Pescara, and Bari (Faričić 2014; Marelić 2016).

After long debate in *zbor* the Grand County of Poljica agreed to dispatch ambassadors to Venice to offer submission in November 1537. This territory passed into the hands of the Venetians not by conquest, but by way of a peaceable negotiation with Andrea Gritti the seventy-seventh *Doge*. The people of Poljica kept alive the feeling of Venetian nationality. *Poglizza* had twelve historic counties and was actually autonomous in home affairs. The counts of *Poglizzani* had rights in the election of *veliki knez*. The word comes from the Slavic Glagolitic script of Poljica—*knez*—equivalent to the Croatian words *knez* and *župan* and to the Venetian *conte*. The certain evidence for its institution comes from a patent letter of 1755 by Marco Bavich preserved in translation in Venice; original documents were written in Cyrillic (Bosančica) under influence of Glagolitic script, and translated by the official *dragomano* from Split for bureaucratic purposes (VEAS b1048: cc. 11–13, 19–21). The counties of Poljica had been established in 1241 (Siebmacher 1873: 14, 53, 54, 94).

This location made the Grand County of Poljica the gateway to and from the Ottoman Turkish Empire and the outer frontier of Venice to the East.

Simultaneously, Venice faced the Turkish menace in the *Stato da Mar* and strove to strengthen the hierarchical and feudal orders in the Dalmatian littoral, in which socially defined roles were assigned impersonally based on categories such as caste and gender, with contractual orders in which individuals could choose their roles and define their terms (Barbarich 1910: 7–9).

Historic or real counties of the extinguished United Kingdom of Hungary, Croatia, Dalmatia, and Bosnia were not disestablished in the Republic of Venice, and had nothing to do with the patriciate of Venice. Since 1586 the right of titles of a sovereignty has been fully established in Venice by legal confirmation or not. The fact is exactly opposite to what is generally supposed to be in a Republic. Families of the Venetian Dominion, by the infeudation of titles to land, had to submit pleas to *Provveditori sopra Feudi* (the commissioners on fiefs), until 1761 (Mosto 1937: I, 184–188; Persichetti Ugolini 1926: 338). In the heraldic system of the French, for instance, a son of the comte-souverain house of De Vismes resides in France; and the comte de Saint Pol deprived himself of the title of sovereign-comte of Neufchatel and Valengin in France (Burke 1867: 1229; Pothier 1839: XIX, 358).

Companies of the Natives of Poljica, the *Poglizzani*, became the most faithful subjects to Venice even at the cost of their lives (VEAS b1048: cc. 27–29). They served in the *Militia*. Lead ore, or, as the Venetians call it, *piombo*, is a vital element in the economy of the *Serenissima* during peace and war. Transportation of lead ore from the mines of the *Stato da Mar* to Venice was granted to the *Poglizzani*, and released on statutory licence; Lorenzo Licini was duly designated as *Pubblico Perito*, i.e. *agrimensore*, in the following document (ZDAZ, Provveditore Generale Pietro Michiel, I, 159, ff. 240–241; f. 241 a-b; Licini 2009: 344). Transportation of lead products by ship from Sebenico (Šibenik), the port of departure, to Venice, the port of arrival, was begun in 1765. Transportation from the port of unloading along Rio de San Zulian to the storage area, which is *Corte Licini o del piombo* going directly by the side entrance and under portico to the Licini's House at Venice, Sestiere Castello, Rio de San Zulian, Calle de la Fava (Fig. 2.5), would be provided under the supervision of the proper public officer, Lorenzo, Licini in Zadar and Venice, and Rubcich in the Grand County of Poljica to denote affiliation (today the name Rubbi stands wrongly in the House complex at Venice as a synonym of Rubcich).[4]

Faithful in peace, and terrible in war, volunteer units from the Dalmatian littoral regularly formed the *Milizia di Mare*, or, as the Senate of Venice called it, *Milizia Oltremarina*, or *Dalmata*, as the units were alternatively called, conducting all peacekeeping operations in support of the Venetian Dominion both in the *Stato da Mar* and the *Stato da Terra*, and often outside it too. The Venetians called them generally *Slavoni*, or *Schiavoni*. Slav units date back to 1519 (Toderini and Checchetti 1873: III, 111 num. 194). What is remarkable about the *Militia di Mare* is how influential it has been in Venetian history and culture since the naval engagement of 1571 which took place in the ports of Corfu, Cefalonia, and Lepanto (West Greece) situated on a bay on the north coast of the Gulf of Corinth, between the *Armata Vinitiana* on the orders of Sebastiano Venier the *Proveditor Generale in Corfu*, and the *Armata Turchesca* on the orders of Mustaffà Bascià. The Turkish fleet entered the Dalmatian *Riviere* in 1571 burning and killing and destroying in their ruthless manner to penetrate deeply into the *golfo Adriatico*. The engagement ended victoriously for the Christians on October 7, 1571. On this day, fifteen thousand Christians and Christian slaves lay dead between Cefalonia and Lepanto, and all the Turkish captains died valiantly (Conti 1589: II, Libro XXII, 152–153, 160–161).

On May 12, 1797, a Militia Corps of 30,000 men with artillery from Dalmatia had filled the Piazzetta and Riva de Slavoni along the water's edge when the *Doge* and patriciate of Venice met in the Hall of the Grand Council in the Ducal Palace for what was to be the last session of its history (Tentori 1799: I, 11–13; II, 383 n. 1, 400–406). Upon the long bank facing the lagoon, in modern times called Riva dei Schiavoni, may be seen a great gate, which leads to a large open space, called St Zaccariah's Field—Campo di San Zaccaria. Then presenting the abdication, the

[4] CVE 2012: Allegato A: *Sestiere Castello*: 46, *sub codice* via: 33160. Tassini's work contains no proof of any scientific knowledge, or any correct notion of the complex (Tassini 1863: I, 245, 319).

Fig. 2.5 Corte named Licini o del piombo (lead ore), Venice, Sestiere Castello, Rio de San Zulian, Calle de la Fava. Detail of the temporary storage rooms. The *Poglizzani* assume monopoly for the transportation of lead ore from Šibenik to Venice (Venice and Zadar, 1765, Provveditore Generale: Pietro Michiel). The family is called Licini in Zadar and Venice, and Rubcich in *Poglizza*

Council was proceeding to deliberate, when an irregular discharge of musketry in the Piazzetta beneath the window filled the Hall with confusion and dismay. Reluctantly, the "Sclavonians" were eventually sent to their native lands across the Gulf of Venice. On the morning of May 16, Venice sent a flotilla to convey to the city a division of 4000 French troops, under General Baraguay d'Hilliers.

4 Conclusion

This investigation has enabled us to emphasize the distinctive character of three manuscript sea charts, and make some deductions about their age and commission; it has also allowed us to recognize these charts as part of a remarkable diffusion of cartographic knowledge of the Adriatic Sea during the century that followed the capture of Constantinople (modern Istanbul) on May 27, 1453.

Mehmed II, who had paused after the capture of Constantinople, resumed his advance, and in 1456 laid siege to Belgrade. Out of this military expansion was created a Turkish empire as a part of the imperial project of Sultan Mehmed II, the Grand Turk, extending from the Bosporus, or, as the Turks called it, the Arm of St George, to the east, to the Adriatic to the west.

Fear then seized the Venetians who dreaded naval assaults on their *Veneto Dominio*. On the eastern shore of the Adriatic, the defence of Istria, Dalmatia, and Albania constituting the beloved Venetian *Stato da Mar* was then undertaken by reinforcing the defences of the system of naval bases and maritime counties in historical Dalmatia. At the same time, the regional system of counties was superimposed by a group of multi-ethnic feudal lords; such jurisdictions were legally recognized by *Provveditori sopra Feudi*, the magistrates concerned with titles of nobility and military fortresses to keep Dalmatia safe from attack. A Militia Dalmata was created to repel invasion, suppress insurrection, defend the Venetian State in war, or operate over different routes in different directions as shown by different-coloured rhumb lines on the portolan chart, thus exemplifying the cultural continuum of the Adriatic area.

Among the Anconitans, the Pope's land, there existed a similar system of maritime counties and trade organization at a very early period of their history. The rich connections between the chartmakers of the Anconitan schools and the Knights of the Order of St Stephen established in 1562 first in Pisa in the Reign of Italy and then in Ancona in the States of the Church in 1563 certainly deserve further study.

Now, we can better understand how Napoleon, in exile, was able to reconsider the affairs of Venice in 1797 in terms of prudence: *Venice was very strong; it was defended by its lagoons, and a great number of armed vessels, while 15,000 Sclavonians formed its garrison. Being mistress of the Adriatic, Venice might still receive additional troops. Concealed inside Venice was the moral strength of all those sovereign families which were about to fight for their political existence. Who could estimate the value of time which our French troops might consume in this new enterprise?*

Thus, the roots of the correct name Slavonians from the word *slava*, meaning glory, become plain.

Primary Sources

Comune di Venezia, Direzione Affari Generali e Supporto Organi (CVE)

Stradario del centro storico veneziano, 27 febbraio 2012, Servizio Ecografico e Toponomastica), Allegato A: *Sestiere Castello*: 46, *sub codice via*: 33160, *specie*: RAMO, *denominazione*: LICINI, *civici*: 5502 5508A 5508B 5516; *scv* 33180, *sp* SOTOPORTEGO *den* LICINI; *scv* 33150, *sp* CORTE *den* LICINI O DEL PIOMBO, *civv* 5503 5503A 5503B 5504 5504A 5505 5507 5508; p. 48, *scv* 35760, *sp* CORTE *den* RUBBI, civv 5509 5510 5511 5511A 5512 5513 5514 5515 5516A.

Loreto, Archivio Storico della Santa Casa (LOASSC)

Libro di spese diverse (1538-1556): c. 32vD in Ancona: adì 2 marzo del 1550 die dar la Comonità de Ancona li regolatori misser Ludovico Gratioli et misser Angelo Freduci [...] per doi bandiere da trombeti; c. 33rD in Ancona: "adì 18 marzo del 1550 die haver la comonità contrascritta fatomi contar el regolator misser Angelo Freduci ali sarti de Loggia per pisone de le loro botege.

Mantova, Biblioteca Comunale Teresiana (MNBCT)

Freducci A (1556): MS 646.

Pisa, Palazzo delle Carovane (PPC)

Sala Azzurra, 450: n. 252, Ioannes Eufredutyus Anconetanus MDLXII (9 marzo 1562, Florentine calendar=1563); n. 266, Franciscus Ferrectus Anconetanus MDLXII (25 dicembre 1562); n. 285, Pirrus Barbolanus ex comitibus Montis Acuti Dns Montis Rotundi MDLXII.

Roma, Biblioteca Nazionale Centrale (RMBNC)

Pareto B (1455): *Carte nautiche 1.*

Roma, Galleria Colonna (RGC)

Bonomini B (1570): *Tabula Nautica.* I am grateful to Patrizia Piergiovanni (Director) and Giacinta Sanfelice di Bagnoli for supplying photographs of Bonomini's chart in the Galleria Colonna.

Trieste, Archivio di Stato (TSAS)

Atti amministrativi dell'Istria (1797-1813), busta 47, carta 650-651: Eccellentissimo Signor Conte Giovanni di Klenau / Cesareo Regio General Maggiore e Comandante del Corpo.

Venezia, Archivio di Stato (VEAS)

Provveditori sopra Feudi – Libro Aureo dei veri Titolati: fascicolo b1048 Licini Rubcich Zara: Terminazione N.o C 815, Magistrato di Zara, 31 agosto 1781 c. 1r, cc. 2-29.
Senato-*Mare*: reg. 15, fol. 14.
Senato, Dispacci, Provveditori da Terra e da Mar e altre cariche, b. 467 (ex 662), Provveditori Generali di San Marco in Dalmazia e in Albania, Andrea Querini, No. 55, 13 febbraio 1797, cc. 4.

Zadar, Državni arhiv u Zadru (ZDAZ)

Atti de Capitani Veneti in Zara dall'anno 1772 al 1777 e dal 1781 in conclusione a tutto l'anno 1797: Provv. Gen. Paolo Boldù, Libro I, carta 63*v*: Il Co[nte] Lorenzo Licini Rubcich, vero titolato del Dominio di Venezia. Addì 18 maggio 1782: Presentate in mano di Su[a] E[ccellenza] Paulo Boldù Provveditore Generale in Dalmazia, ed Albania.

Indice degli atti de' Capitani ex Veneti in Zara: Pietro Orio, Capitanio di Zara dall'anno 1781 all'anno 1783, Libro I, Co: Licini Rubcich Lorenzo, carta 455, 456, 457, 459; *Marchio Querini, Capitano di Zara dall'anno 1783 all'anno 1785*, Libro I, Licini Lorenzo Rubcich, carta 358; Licini Lorenzo Rubcich, carta 380, 381; Licini Domenico Rubcich, carta 474, 480. *Atti de' Provveditori Generali ex veneti in Dalmazia ed Albania, Provveditori Generali, Gio. Battista Loredan Diedo*, Zara, 10 giugno 1751: Decreto per Poglizani. *Atti de' Provveditori Generali ex veneti in Dalmazia ed Albania, Provveditori Generali Pietro Michiel 1763-1765*, Zara, Libro I, carta 159, ff. 240-241: Monopolio del piombo ai fedelissimi Poglizzani, Ducale datata Venezia 25 maggio 1765, trascritta in Zara, 28 luglio 1765; f. 241 a-b: Elezione del Conte Lorenzo Licini in Pubblico Perito di questa Provincia, 30 luglio 1765.

Appendix

ZDAZ, *Atti de' Provveditori Generali ex veneti in Dalmazia ed Albania, Provveditori Generali Pietro Michiel 1763–1765*, book I, box 159, ff. 240–241 a-b, Monopolio del piombo ai fedelissimi Poglizzani; f. 241 a-b, Elezione del Conte Lorenzo Licini in Pubblico Perito di questa Provincia.

Ducale per l'Appalto de Piombi di questa Provincia—Alli 28 Luglio 1765. Zara. Prex[en]tate: dal Sig[nor] Lorenzo Ruste per nome dell'infrascritto Alberto Bironi nelle mani / di S[ua] E[ccellenza] Pietro Michiel Proc[uratore] G[enera]le in Dalma[tia], et Alba[nia], da cui vedute, et lette, fù ordinato il di loro registro, et pontuale esecuz[ion]e. *Aloÿsius Mocenico* Dei Gratia *Dux Venetiarum c[um] Nobili, et Sapienti Viro Petro Michaeli Provisori N[ost]ri G[enera]lis in Dalma[tia], et Alba[nia]*: Fideli Dilecto salutem, et dilectionis affectum. Dalla unita deliberazione rilevarà la virtù V[ost]ra: essere stata accettata dal Senato la offerta, che vi si occlude, proddotta dall'Int[ermediari]o Filosi per conto del suo principale Bironi, per l'affittanza del Partito de Piombi grezzi, et Ballini della Provincia, et Isole Superiori, coll'aumento del 105 per 100, e con quello pure del 30 p[e]r cento sopra il consumo delle libre m/16 [16 migliara] piombo per anni venti, avendo assentito alle regolazioni tutte della Polizza d'Incanto, che vi sarà trasmessa dal Mag[istrat]o de Dep[uta]ti, et Agg[iun]ti, come consta dalla copia dell'annesso Costituto. Cristate [firmate con nome e cognome], che abbia il d[ett]o Bironi le necessarie pieggierie, ordinarà il zelo vostro, che sia posto al possesso d'esso Prestito, e piantato debitore di annue lire seimille cento cinquanta di cor[ren]te regolata moneta, che doverà contribuire in due eguali Ratte, à tenor del suo impegno, e praticare l'annuo consumo di piombo provisto alla Dominante in summa di libre ventimilla ottocento. Da[t]o in No[st]ro D[uca]le Palatio Die XXV Maij In[dicion]e XIII, MDCCLXV. Gio: Manin Seg[retar]io.

Elezzione del Co[nte] Lorenzo Licini in Pub[blic]o Perito di questa Provincia—I[llustrissi]mi S[igno]ri. Munito di tutti li requisiti voluti dalla Determinazione dell'Eccellentissimo Signore Francesco Grimani Predecessor Nostro il 22 Giugno 1756 si è rassegnato alla carica il Co[nte] Lorenzo Licini abitante in questa Città, supplicando d'essere ascritto frà li Publici Periti, onde poter legalmente fungerne l'incombenza in questa Provincia.

Esaminati da Noi li requisiti sudetti, da quali risultandoci la sufficienza, e puntualità del medesimo Licini, non menoche l'onoratezza de di lui costumi, e che non sia stato mai processato, siamo perciò concorsi ad esaudirlo, lo più coll'autorità dal Stato Nostro, et in virtù delle presenti eleggiamo, et destiniamo il sup[radett]o Conte Lorenzo Licini in Publico Perito di questa Provincia, con tutti obblighi annessi, le prerogative, e ricognizioni spettanti al detto Impiego, esercitati, et godute rispettivamente da tutti gl'altri Publici Periti della Provincia, et dichiariti nella Determinazione suaccennata, copia della quale anderà debito esso Licini di levare da questo Publico Archivio, onde serva di norma impreteribile all'esercizio di sue incombenze. Per tale adunque dovrà essere riconosciuto, et nel Publico Servizio rispettato, ordinando delle presenti il registro ove occorre, per la loro esecuzione. Zara li 30 Luglio 1765 Pietro Michiel.

References

Agostini F (1998) L'Area alto-adriatica dal riformismo veneziano all'età napoleonica, Venezia, Marsilio, Istituto per le ricerche di storia sociale e religiosa, istituto veneto di scienze, lettere ed arti

Almagià R (1944) Monumenta Cartographica Vaticana, Città del Vaticano, Biblioteca Apostolica Vaticana, 1944, vol 1, xi–156.

Altić M (2016) Between Secrecy and Silent Cooperation: The Dissemination of Knowledge about the Republic of Dubrovnik in the Context of the Ottoman-Venetian and Napoleonic Wars. In: Altić M, Demhardt IJ, Vervust S, Cham Springer IP (eds) Dissemination of cartographic knowledge: 6th international symposium of the ICA commission on the history of cartography. Lecture Notes in Geoinformation and Cartography, 55–74

Altić M (2023) Cultural and knowledge exchange between Dubrovnik and Livorno at the Time of Ferdinando I de' Medici through the Eyes of Vincentius Demetrius Volcius Raguseus. In: Storia Militare Moderna, Fascicolo 15, Anno 4 (giugno 2023), (Ed. Virgilio Ilari), 41–82

Astengo C (2007) The renaissance chart tradition in the mediterranean. In: Woodward D (ed) The history of cartography, vol 3. The University of Chicago Press, Chicago, Part 1, The History of Renaissance Cartography: Interpretive Essays, Essay 7, 174–262.

Baldini B (1578) Vita di Cosimo Medici. Primo Gran Duca di Toscana, Firenze, Bartolomeo Sermartelli

Barbarich E (1910) La Campagna del 1796 nel Veneto., Roma, Enrico Voghera Editore, 1910

Belgrano LT (1866) Correzioni ed aggiunte relative a la cartografia ligustica, u: Atti della Società Ligure di Storia Patria, IV, 491–496

Biemmi G (1756) Istoria di Giorgio Castrioto detto Scanderbeg-Begh, Brescia, Giammaria Rizzardi

Burke B (1867) Genealogical and heraldic dictionary of the peerage. Harrison, London

Campbell T (1987) Portolan charts from the Late Thirteenth Century to 1500. In: Harley JB, Woodward D (eds) The history of cartography, vol 1: cartography in prehistoric, ancient and Medieval Europe and the Mediterranean. University of Chicago Press, Chicago, pp 371–463

Casey ES (2002) Representing place: landscape painting and maps. University of Minnesota Press, Minneapolis and London

Ciavarini C (1867) Sommario della storia di Ancona raccontata al popolo anconitano, Ancona-Como, A. Giorgetti

Conti N (1589) Delle historie de' suoi tempi, Venezia, Damian Zenaro, Parte Seconda

Dall'Ara F (2020) Sant'Agata dei Fregoso, Sant'Agata, Dall'Ara Ed., II

Dandolo G (1855) La caduta della Repubblica di Venezia ed i suoi ultimi cinquant'anni. Venezia, Pietro Naratovich:1

D'Israeli I (1793) Curiosities of literature. Consisting of anecdotes, characters, sketches, and observations, London, Murray

Faričić J (2014) The significance of the Croatian Coastline in the Network of European Pilgrim Routes. In: Katić M, Klarin T, McDonald M (eds) Pilgrimage and sacred places in Southeast Europe: history, religious tourism and contemporary trends. Lit Verlag, Berlin, pp 25–47

Ferretti IB (1566) Constitutiones, sive Statuta magnificae civitatis Anconae, novissime decreto Senatus impressa, Ancona, Excudebat Astulfus de Grandis Veronensis

Ferretti F (1580) Diporti Notturni. Dialoghi Familiari del Cap.o Franc.o Ferre.ti Cav.ro dell'Ordine di Sa.to Stefano. Con la Dimostratione Figurale Intagliata da Michel'Angelo Marrelli Anconitano 1579, Ancona, Salvioni, Notturno Secondo

Ferretti F Jr (1685) Pietra del paragone della vera nobiltà discorso genealogico de conti Ferretti con varie notizie historiche, e riflessioni sopra i pregi della Nobiltà. Alla Maestà Christianissima di Luigi XIV, Ancona, Serafini

Foscarini M (1722) Degl'istorici delle cose veneziane, i quali hanno scritto per Pubblico Decreto, Tomo Decimo che comprende l'Istoria della Repubblica Veneta di Michele Foscarini Senatore, Appresso il Lovisa, Venezia, Libro Primo

Caspar E (1923) Gregorii vii Registrum. Lib. V-IX. Das Register Gregors VII. II, Buch V-IX, u: MGH, Epistolae selectae, Tomus II, Fasciculus II, Berlin, 348–711, Liber VIII, 1 b, Investitur Robert Guiscards durch Gregor VII. Ceprano, 1080, Juni 6, 515–516

Guglielmotti A (1862) Marcantonio Colonna alla Battaglia di Lepanto, Firenze, Felice Le Monnier

Guglielmotti A (1871) Storia della Marina Pontificia nel Medio Evo dal 728 al 1499, Firenze, Successori Le Monier, I, Libro I

Guglielmotti A (1886) La guerra dei pirati e la Marina Pontificia dal 1500 al 1560, Roma, Vaticana, Volume I

Johnson KG (2016) Napoleon's War at Sea, u: Napoleon and the Operational Art of War: Essays in Honor of Donald D. Horward (Ed. Leggere MV), Lieden, Boston, Brill, History of Warfare / 110, 387–475

Kolumbić J (1979) Grbovi zadarskih plemićkih obitelji. Radovi Zavoda za povijesne znanosti HAZU u Zadru, Zadar 26:55–119

Kozličić M (1995) Monumenta Cartographica Maris Adriatici Croatici. Zagreb, AGM

Medici C (1562) Statuti Capitoli et Constitutioni del Ordine di Santo Stefano. Fondato et dotato dall'Illustr. et Eccell. Signor Cosimo Medici Duca II. di Fiorenza et di Siena, Firenze, L. Torrentino

Novak D, Lapaine M, Mlinarić D (2005) Pet stoljeća geografskih i pomorskih karata Hrvatske / Five Centuries of Maps and Charts of Croatia, Zagreb, Školska knjiga

Las Cases, EADMJ (1823) Mémorial de Sainte-Hélène, ou Journal où se trouve consigné jour par jour, ce qu'a dit et fait Napoléon durant dix-huit mois. Par le comte de Las Cases, Bruxelles, H. Remy, Tome Quatrième

Licini P (2009) Un cartografo nella Dalmazia veneziana: Lorenzo Licini Conte di Poglizza (1725-1802). In: Lago L, Selva O, Umek D (eds) La Toponomastica in Istria, Fiume e Dalmazia. Firenze, Edizioni IGM, 2009, II: Aspetti cartografici e comparazione geostorica, Parte Prima (Aspetti conoscitivi), pp 247–272

Licini de Romagnoli P (2020) Count Freducci's Nautical Charts, Papal Cartography, and Transmission: A Reply to Article by Chet Van Duzer, In:

Humanities, Anno IX, Numero 17 (giugno 2020), Università di Messina, pp 29–144
Lučić I (1666) Ioannis Lucii Dalmatini De regno Dalmatiae et Croatiae libri sex, Amstelaedami, Apud Ioannem Blaev, 1666, Lib. I, Capp. XII–XIII
Marelić T (2016) Wind Influence on Sailing Ship Navigation across Croatian Part of Adriatic Sea. Geoadria 21(2):211–236. https://doi.org/10.15291/geoadria.17
Mosto (da) A (1937) L'archivio di Stato di Venezia. Indice generale, Storico, Descrittivo ed Analitico con il concorso dei Funzionari dell'Archivio, Roma, Biblioteca dell'Arte Editrice, Annales Institutorum Vol. V, Tomo I, Archivi dell'Amministrazione Centrale della Repubblica Veneta e Archivi Notarili: Provveditori sopra Feudi, Libro aureo dei titolati
Orbini M (1601) Il Regno de gli Slavi Hoggi Corrottamente detti Schiavoni: Historia, Pesaro, Girolamo Concordia
Pavich von Pfanenhtal A (1907) Beiträge zur Geschichte der Republik Poljica bei Spalato, in Wissenschaftliche Mitteilungen aus Bosnien und der Herzegowina, Wien, Archäologie und Geschichte, I, Bd. X, p. 70, 308, 311/3
Persichetti Ugolini E (1926) I titoli nobiliari nella Serenissima Repubblica di San Marco, Rivista del Collegio Araldico, Anno XXIV, N. 9 (settembre 1926), Roma, Presso il Collegio Araldico, 385–390
Pothier RJ (1839) A treatise on the law of obligations or contracts. Translated from the French, Philadelphia, Robert Small, II
Rajaković M, Kljajić I, Lapaine M (2014) Map Projection Reconstruction of a Map by Mercator, Cartography from Pole to Pole. In: Buchroithner M, Prechtel N, Burghardt D (eds) Selected Contributions to the XXVIth International Conference of the ICA, Dresden 2013, Berlin, Heidelberg, Springer, pp 31–45
Reuter T, Nelson JL (2006) Medieval polities and modern mentalities. Cambridge University Press, Cambridge and New York
Siebmacher J (1873) Grosses und Allgemeines Wappenbuch: IV, Wappenbuchern 1605-1961. Der Adel des Kőnigreichs Dalmatien, Nürnberg, Band 4, Abthg. 3, tav. 9, riga 3/g3; tav. 35, riga 2|1, pp. 14, 53, 54, 94
Tassini G (1863) Curiosità veneziane, Ovvero origini delle denominazioni stradali di Venezia, Venezia, Grimaldo. Prima Edizione I
Tentori C (1799) Raccolta cronologico-ragionata di documenti inediti che formano la storia diplomatica della rivoluzione e caduta della Repubblica di Venezia, Augusta (ma Venezia), s. e
Toderini T, Checchetti B (1873) Il R. Archivio Generale di Venezia, Venezia, Pietro Naratovich, III. Note Statistiche degli Archivi Veneti Antichi, Collegio della Milizia da Mar, sec. XVI-XIX (1519-1806), fascicoli e registri 688
Weiland L (1893) Constitutiones et Acta Publica Imperatorum et Regum. Tomus I Inde ab A. DCCCCXI usque ad A. MCXCVII, u MGH, Legum sectio IV,

Hannover, 1893: Nr. 390, Nr. 391: Gregorii VII. Concilium Romanum. 1080. Mart. 7., 554/390; Propositio Rudolfi Regis et Principum Imperii, 556/391; Gesta Concilii. Gregorii vii Concilium Romanum, 1. Sequentes statuta sanctorum patrum, [...], A.D. 1080, Nr. 391, 4

CHAPTER 3

Early Modern Nautical Charts of the Adriatic Sea as a Medium of Communication

Julijan Sutlović

1 Introduction

The Adriatic Sea, historically and during the early modern period, was a significant maritime corridor in the Mediterranean. Its strategic position as a connection between continents made it an important route for intense economic and cultural exchanges (Braudel 1997). Coastal settlements, ports, and a network of islands provided opportunities for the development of maritime trade while presenting significant navigational challenges. Nautical charts were indispensable for ensuring safe navigation in this complex geographic landscape, but their purpose extended beyond mere technical aids.

Communication can be simply defined as "the sharing of experiences." People communicate using various signs, including symbols, which enable indirect and mediated sharing of experiences. Symbols can be defined as entities that represent or refer to something else (Tubbs 2003). They

J. Sutlović (✉)
Department of Geography, University of Zadar, Zadar, Croatia
e-mail: jsutlovic21@unizd.hr

J. Faričić, T. Marelić (eds.), *Early Modern Nautical Charts of the Adriatic Sea*, https://doi.org/10.1007/978-981-96-8046-7_3

convey complex messages that go beyond the graphical (visual) or linguistic simplifications of the subject or element involved in communication. This symbolic dimension is evident in nautical charts, which combined practical navigation with symbolic representation to convey a deeper understanding of the Adriatic region.

In 2021, members of the International Cartographic Association adopted a new definition of a map: "A map is a medium designed for the communication of generalized spatial information and their relationships" (Lapaine et al. 2021: 1). This concise definition directly identifies maps as media for communication. Early modern nautical charts of the Adriatic align with this definition, serving as tools for both navigation and the dissemination of cultural, political, and spatial knowledge. Koláčný's (1969, 1977) communication of cartographic information model provides a valuable framework for understanding the interactive role of charts: they acted as a medium for transmitting encoded information from cartographers (senders) to chart users (receivers), with successful decoding relying on the latter's knowledge and experience.

According to John B. Harley (1988), maps are not objective representations of space, but social constructs shaped by the perspectives of their creators. Nautical charts of the Adriatic epitomize this concept, encoding complex layers of meaning through their elements. The studied charts were part of a broader system of spatial organization and communication. Cartographic signs, toponyms, scales, allegorical representations, and symbols were used to encode information tailored to the needs of users and patrons alike. This encoding process involved choices by cartographers, shaped by the technical demands of navigation and maritime powers.

Nautical charts reflected the contested nature of the Adriatic, where the Venetian Republic, the Ottoman Empire, and the Habsburg Monarchy competed for control. These powers, particularly the Venetians, used charts to assert dominance, legitimizing their territorial claims and shaping perceptions of the region's geography. For instance, the Venetian naming of the Adriatic Sea as the *Golfo di Venezia* signified political hegemony. Similarly, religious and confessional symbolism, such as the cross or the crescent moon, depicted on compass roses, underscored the ideological divide between Catholic and Islamic powers (Faričić et al. 2023a). This research investigates the multifaceted roles of charts, emphasizing how early modern nautical charts of the Adriatic were dynamic tools of communication.

2 Research Methodology

The research employed a combination of qualitative and quantitative methods, with a predominance of qualitative approaches. Key methods included the analysis of relevant literature to contextualize the problem and establish a foundation for the comparative study of early modern nautical charts of the Adriatic Sea. The analytical-interpretative method was utilized within a historical-geographical framework to conceptualize new insights. The diachronic method facilitated the tracking of the development of nautical charts over centuries and the analysis of changes in content quality, design, and communicative potential.

The research aimed to investigate early modern nautical charts of the Adriatic Sea as vital communication media, emphasizing the decoding of cartographic elements. It concentrated on understanding how charts, through their representation of geographic reality, cultural and political influences, and navigational needs, conveyed complex messages that extended beyond the technical aspects of navigation. Special emphasis was placed on the analysis of symbolism, toponyms, cartographic signs, and scales to clarify how these charts reflected and shaped perceptions of the Adriatic Sea.

The theoretical framework was grounded in communication theory, interpreting nautical charts as multifaceted media for transmitting information between cartographers (senders) and chart users (receivers). These charts were considered complex communication channels, conveying not only geographical information but also political, cultural, and ideological messages. Deconstruction analysis, inspired by the works of Harley (1988, 1992), had an important role in uncovering the connections between the political, religious, and social contexts of chart production and their content.

Toponym analysis involved comparing geographic names from late medieval, early modern, and contemporary nautical charts to evaluate their significance for navigation and socio-economic functions. The study also examined cartographic signs crucial for navigation, focusing on those typical of nautical cartography. Significant navigational signs on the studied charts were highlighted, and a comparative analysis was conducted, comparing these signs with those used on modern nautical charts. Additionally, the analysis of scales documented the number of graphical scales, and the types of units of distance used across charts. Communication noise was identified through recurring errors across different charts, further enriching the understanding of cartographic communication practices.

The research encompassed 84 nautical charts of the Adriatic Sea (42 manuscript and 42 printed), predominantly created in Venice, the leading maritime centre of the Adriatic at the time. Complementary to chart analysis, sailing directions provided additional context. Charts were sourced from institutions such as the Biblioteca Nazionale Marciana in Venice, the Croatian State Archives, the National and University Library in Zagreb, and the University of Trieste. Digital reproductions from European and American libraries and private collections, particularly the Marco Asta collection, were also utilized.

3 The Complexity of Maps as a Medium of Communication

Understanding the complexity of maps as a medium of communication requires a concise historical perspective on the evolution of cartographic theories. By the late nineteenth century, cartographers sought to establish cartography as a science. A key concept of this period was the "logic of maps," as defined by Max Eckert (1908). He argued that scientific rigour would eliminate subjectivity and ensure the objectivity of maps, despite their artistic elements. Eckert described maps as products of "art clarified by science." While he attempted to establish a theoretical framework for cartographic practice, the lack of advanced communication and semiotic theories at the time limited his practical impact (Pápay 2018). However, Eckert's work laid the foundation for the functionalist theory that emerged after World War II.

In 1952, Arthur H. Robinson's work *The Look of Maps* established the foundations of the communication paradigm in cartography, defining maps as a medium for transferring information between cartographers and users. His concept of "functional design" emphasized the importance of visual methods for clear and effective communication. Robinson stated, "the most scientific cartography concerns the dissemination of spatial knowledge" (Robinson 1952: 17), implying that maps are primarily tools for conveying the author's spatial understanding without considering user needs or feedback.

3.1 Communication Theory in Cartography

Within communication theory, the central issue is the accurate reproduction of the intended message across stages of the communication process. Maps are conceived as representations of reality, designed to provide users with precise and readable information. However, their success depends on understanding users' needs, knowledge, and context of use (Koláčný 1977). Claude Shannon and Warren Weaver's (1949) foundational communication model, which includes an information source, sender, channel, receiver, and destination, was later applied to cartography. Models simplify reality to highlight aspects crucial for solving specific problems (Maletzke 1998). In cartography the transmission of information occurs within the system of cartographer–map user, underscoring the importance of understanding the essential features of cartographic communication (Robinson and Petchenik 1975).

Maps differ from other media as they require specific skills to interpret and understand spatial relationships. Cognitive processes can lead users to interpret the same map information differently, highlighting the importance of effective design and standardized conventions. Konstantin A. Salitchev (1970) emphasized that cartography has strong ties to the theory of scientific information, viewing maps as critical sources of knowledge. Michael K. Bocharov (1967, as cited in Salitchev 1970) defined cartography as the science of transmitting information through cartographic forms, linking it with information sciences. Salitchev further emphasized the user's role through the concept of "theory and methods of map use," complementing the basic communication model.

Antonín Koláčný (1969) defined communication of cartographic information as a unified process that integrates the creation and use of maps. This approach enables the development of maps that meet not only technical standards but also the actual needs of users, ensuring optimal functionality in diverse contexts. His communication model focuses on cartographic information rather than the map's content. Cartographic information encompasses the intrinsic meaning and interpretation of the reality represented by maps. He outlined seven stages in his model, from selective observation of reality to intellectual transformation and eventual user application. This process emphasizes the selection, transformation, and objectification of information to create cartographic symbols that are comprehensible to users (Fig. 3.1). It begins with the cartographer's selective observation of reality, transforming a multidimensional model into a

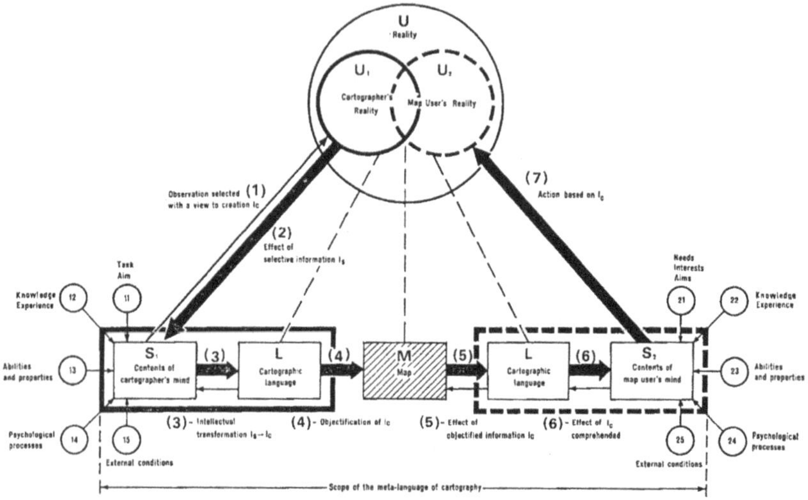

Fig. 3.1 Koláčnýs's communication of cartographic information model (Source: Koláčný 1969: 41)

two-dimensional representation. The information provided by the map enriches the user's knowledge, allowing for a reinterpretation of reality based on newly acquired cartographic insights. This process underscores the importance of not only technical precision but also a profound understanding of user needs, perception, and cognitive abilities (Koláčný 1969, 1977).

The aforementioned communication model provided a comprehensive view of the interaction between cartographers and users, highlighting that maps are not passive media but essential elements in the meaning-making process. The communication of cartographic information has become a key component of cartographic paradigms, integrating map creation and use into a unified process. This integration improves the functionality and precision of maps, ensuring they meet user requirements. In cartographic communication systems, key concepts include the observer, the map user, and the map reader (Robinson and Petchenik 1975). Observers acquire spatial knowledge through map interpretation, while users and readers utilize maps for specific purposes, such as determining navigational routes, without necessarily expanding their knowledge.

3.2 *The Layered Dimensions of Maps as Medium of Communication*

According to John S. Keates (1984), cartography is both a scientific and artistic discipline. While scientific approaches ensure accuracy and functionality, excluding art from cartography diminishes its aesthetic and symbolic value. Maps often include artistic elements, such as textures, shading, or iconographic symbols, that enhance their interpretive richness. Maps cannot be reduced to a binary division of scientific and artistic elements; their value lies in the complementarity of these approaches. The scientific approach employs reductionist methods to analyse processes, while the artistic approach provides a holistic perspective through intuition and critical experience. These approaches are mutually enriching, particularly in aspects such as perception, colour theory, and visual aesthetics (MacEachren 2004).

Postmodernist theorists such as Denis Wood and John Fels (1986), Wood (1992), and Harley (1988, 1992) highlighted the inherent subjectivity and rhetorical content of maps, opening new dimensions in cartographic studies. They emphasized the role of maps as social and political constructs. Harley's deconstructionist approach analysed maps as textual constructs imbued with symbolism, politics, and ideology. Maps are not neutral representations of reality; they legitimize dominant narratives, control space, and shape social reality. For instance, symbolic elements such as flags and allegorical depictions on early modern charts often served as tools of political legitimization, conveying messages of territorial aspirations and hegemonic ideologies.

A semiotic approach views maps as systems of signs with multiple meanings. According to Alan M. MacEachren (2004), the meaning of cartographic symbols is divided into denotative (what the map represents) and connotative (its broader meaning and symbolism). For example, while blue on maps often denotes water, light blue can signify shallow areas on nautical charts. Connotatively, symbols may include political or cultural implications, such as boundaries reflecting territorial ambitions.

Postmodernist perspectives, including discursive analyses (Barnes and Gregory 1997), argue that cartographic representations are not neutral but reflect social values and ideologies. Maps are not merely depictions of space but active participants in shaping it (Pickles 2004). Rob Kitchin and Martin Dodge (2007) proposed an ontogenetic approach, viewing maps as dynamic processes that transform with each use. Maps are not static objects but tools for solving spatial problems in specific contexts. Each user interprets maps based on their needs, prior knowledge, and cultural background, making maps relational and mutable entities. The post-representational approach focuses on the social implications of cartographic representations, enabling a deeper understanding of the role maps play in coding the world.

A study of these theoretical frameworks reveals that maps are multi-layered media, integrating scientific, artistic, semiotic, cultural, and other dimensions. Their interpretation requires integrating various theoretical frameworks, from functionalist and semiotic models to postmodernist critiques. Early modern nautical charts exemplify this complexity, where decorative elements, toponyms, cartographic signs, and other chart elements not only convey spatial information but also reflect the social, political, and cultural power dynamics of their time.

4 Imagology: Compressing Reality into Symbolism

The effectiveness of maps as communication tools lies in their persuasive and suggestive power, often causing users to disregard possible distortions of reality. As semiotic systems, maps contain symbolic representations of compressed reality, organizing geographical knowledge into visual schemes subject to culturally accepted interpretations of territories—both real and abstract. Their communicative potential also includes an imagological aspect, whose interpretation hinges on the dichotomy of Self/Us and the Other (Culcasi 2008). Geographically, the Other is placed on the margins, reflecting the discursive construction of identity and dominance, often tied to imperial and colonial projects (Mountz 2009).

The deconstruction of maps, as proposed by Harley (1988, 1992), is important for uncovering the ideologies and power structures that shaped their content. Identifying political, religious, or social centres of power reveals how cartographers depicted their empires as superior, often excluding the Other from charted spaces (Mlinarić and Gregurović 2011). An example of this is the naming of the Adriatic Sea as the Venetian Gulf (*Golfo di Venezia*), portraying the entire Adriatic as an extension of the Venetian harbour (Marković 1993; Faričić et al. 2023b). Early modern nautical charts, like other maps, allowed users to perceive spaces through the lens of imperial and cartographic policies, while providing a historical and cultural context. The messages conveyed by charts can also be interpreted through vexillological, heraldic, and allegorical elements, which often reflect the political, cultural, and confessional narratives of their time.

4.1 Flags and Heraldry: Communicating Territorial Authority and Identity

Late medieval and early modern cartographers used flags and heraldic symbols on charts to convey layered messages about territorial authority, religious identity, and political relations. Examples such as the charts of Angelino Dulceti (1325/1330), Antonio Millo (1583), Joan Martines (second half of the sixteenth century), and Pietro Cavallini (1677) illustrate how these symbols functioned not only as aesthetic elements but also as tools for decoding territorial and cultural relationships, even in the absence of delineated borders.

On the eastern Adriatic coast, Dulceti depicted three flags: a quartered blue-and-white flag near the toponym *Narent*, a rectangular yellow flag with a white disc and red cross near *Sibinicho*, and a red-and-yellow flag with a white star near *Segna*. These flags marked territories under the jurisdiction of various entities. Dulceti also included flags bearing symbols such as a crescent moon (associated with Islam and Arabian sultanates), the winged lion of St Mark (symbolizing Venice), the stylized lily (symbolizing Florence), and two crossed keys (representing the Papal States), among others. These features indicated territorial control and confessional identities, reflecting the Adriatic as a site of long-standing imperial conflicts and encounters between diverse cultures.

Another example is Millo's 1583 chart, which differentiates the territorial and confessional affiliations of the Adriatic coast through nine flags adorned with heraldic symbols, including the winged lion, double-headed eagle, crescent moon, and St Blaise (Fig. 3.2). The largest flag, depicting the winged lion of Venice, is placed near Venice, with smaller ones positioned near Split (*Spalato*) and Kotor (*Cataro*). A square flag with a black double-headed eagle on a gold field, located near Rijeka (*Fiume*), represents the Habsburgs—kings of Hungary and Croatia and emperors of the Holy Roman Empire. Flags featuring crescent moons, symbolizing Islam and the Ottoman Empire, are placed near towns such as Obrovac (*Brouazo*), Drijeva or Gabela (*Naronra*), Durrës (*Durazo*), and Butrint (*Butrintio*). The flag of St Blaise, the patron saint of Dubrovnik (*Ragusi*), is placed near that city. Millo also depicted two coats of arms: the Keys of

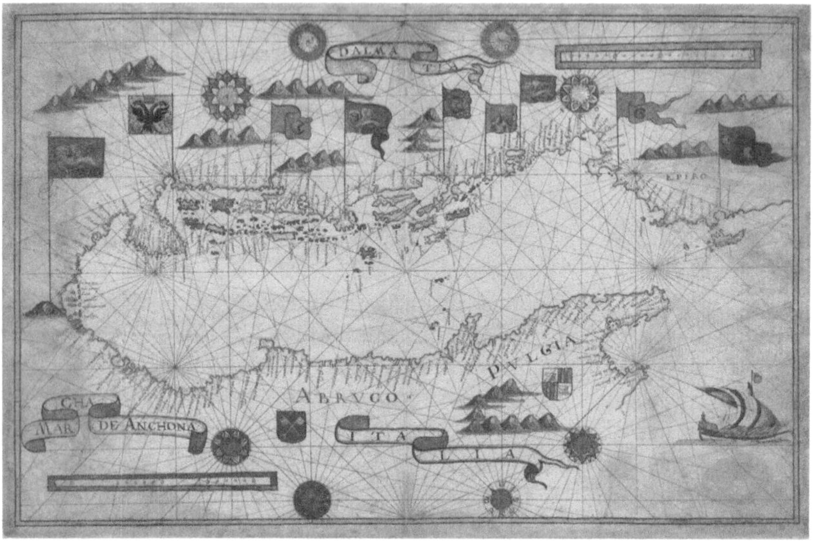

Fig. 3.2 Antonio Millo, nautical chart of the Adriatic Sea, Venice, 1583 (Source: Biblioteka Narodnowa, Warsaw, BN ZZK 0.2399)

Heaven on the coast of Abruzzo, representing the Papal States, and two diagonally positioned towers on the coast of Apulia, representing the Kingdom of Naples.

Vexillological and heraldic practices on late medieval and early modern nautical charts highlight how visual symbols were leveraged to communicate authority, territory, identity, religion, and cultural affiliations, extending their meaning beyond mere decoration. These elements linked charted spaces to specific communities or cultural spheres, providing insights into the narratives of their time.

4.2 *Metaphors of Dominion: The Allegorical Venice*

Allegories have appeared on maps since antiquity. On Venetian charts, allegories became increasingly common, particularly with the advent of printed charts. Venetian political culture in the early modern period—perhaps more than any other city—relied on visual imagination to shape and communicate its political ideology, and allegories to give a pictorial form to their sovereignty (Horodowich 2021).

One of the most frequent allegorical depictions on charts is the Venetian lion—lion of St Mark. Beyond charts, this symbol appears on Adriatic coastlines under Venetian rule—in reliefs, statues, paintings, and written works. That lion is characterized by its saint's (in fact Evangelist Mark's) halo, book, and wings. Most lions are depicted as *leone alato andante*—a walking lion shown in profile. Occasionally, the lion is portrayed in a unique Venetian pose, seen from the front with spread wings, called *leone in moleca* (or *leone in maesta*), resembling a crab with open claws. This specific iconography symbolized Venice's divine mandate, its expansion from sea to land, and the sacred unity of Venice, its lagoon, and the Adriatic Sea (Zorzi 1980; Griffith 2005; Pedani 2006; Rizzi 2012).

Alvise Gramolin's 1624 chart features the winged lion of St Mark depicted as *leone alato andante*. In a similar vein, Jean François Roussin's (1661) chart features a winged lion positioned above a Venetian cityscape, visualy dominating the chart and underscoring Venetian propaganda and dominance over the depicted region. The chart's title, *Carta dv Golfo di Venetia*, further emphasizes Venetian claims over the Adriatic.

On charts by Gasparo Tentivo (circa 1700) and Vincenzo De Lucio (circa 1792–1796), nearly identical personifications of Venice appear (Fig. 3.3). Tentivo titled his chart *Carta Marittima del Golfo di Venezia con tutte l'Isole e Scogli che in esso si trovano* (Maritime Chart of the Venetian Gulf with All Islands and Islets Found Within It). By inscribing the title across the horonym *Dalmazia*, Tentivo visually suggested Venetian control over this part of the eastern Adriatic coast. De Lucio incorporated an allegorical depiction of Venice on both the chart sheet representing Central Dalmatia and the one depicting the island of Zakynthos (*I. Del Zante*) in the Ionian Sea—one of Venice's southernmost possessions,

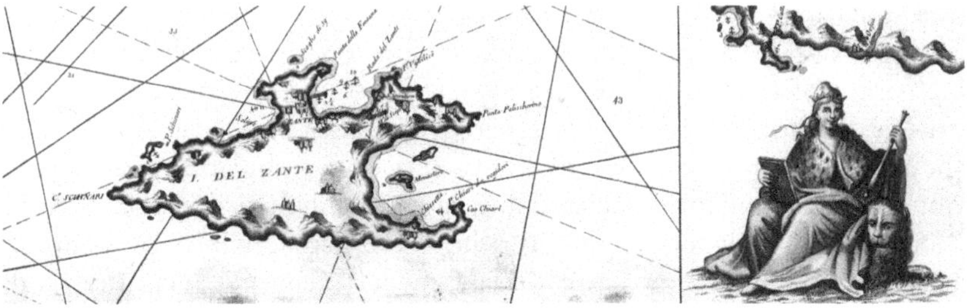

Fig. 3.3 The personification of Venice; Vincenzo de Lucio, ca. 1792–1796. (Source: Private collection Marco Asta, Bologna)

intermittently ruled from 1484 to 1797 (Miller 1903)—further asserting its imperial narrative.

Venice was often personified as a female figure. Tentivos's and De Lucio's allegories portray a richly adorned woman draped in a sable fur cloak, wearing the *Corno Ducale* (doge's cap), holding a book in her right hand and a sceptre in her left—symbols of the doge's authority (Garret 2001). The lion of St Mark is placed beside her, reinforcing Venice's dominion. Venice was also called the "Queen of the Adriatic" (Davenport Adams 1869), a title reflected in numerous books, paintings, and other works of art.

Allegorical representations on early modern nautical charts caption the sophisticated visual imagination of Venetian cartographers in articulating political ideology and territorial sovereignty. Venice employed a wealth of allegories to create powerful visual representations of its cardinal virtues and political culture. Allegories, as metaphorical forms, enabled users to recognize and identify deeper meanings in symbols and personifications, pointing to the political aspirations and cultural authority of the Venetian Republic.

5 Toponyms: Navigational, Cultural, and Political Dimensions

A toponym is a concise linguistic descriptor that effectively replaces excessive descriptions, facilitating spatial orientation and everyday communication (Faričić 2011). Nautical charts, being more than technical instruments, function as expressions (and productions) of ideological values, social and cultural norms, and worldviews. In this context, toponyms, inter alia, have a crucial role in understanding the performative process of "territorialization" and can be used to study cultural identity and political turmoil in the charted space. Changes of toponyms are not only part of the historical fluctuation of language, but they also act as "markers" of geographical features, territories, and identity spaces (Siniscalchi and Palagiano 2018). They are the longest-lasting link between people and places, providing a robust foundation for understanding cultural interaction (Skok 1950). Their analysis provides insights into how geographic names on nautical charts reflect and shape navigational practices, cultural identities, and political territories.

5.1 Hierarchy and Style in Toponyms

The characteristics of toponyms differ significantly between manuscript and printed charts, with the style of writing toponyms being a distinctive feature of portolan charts. On these charts, toponyms are most often inscribed perpendicular to the coastline, creating a curvilinear effect that follows the shoreline. Toponyms for islands, inscribed within the marine spaces of the chart, are typically rotated 180° compared to those representing mainland locations. According to Harley (1992), hierarchical tendencies in cartography involve omitting less significant spatial features—those outside the immediate purpose of the chart—in order to emphasize key information. This is evident in the hierarchical structuring of toponyms on early modern nautical charts of the Adriatic. On manuscript charts, more significant toponyms, such as those denoting safe harbours with resources like water and provisions, were marked in red (Nordenskiöld 1897; Brown 1979; Monmonier 2004), while others were written in black. With the advent of printing, the practice of marking toponyms in red and rotating island names ceased. All toponyms began to be inscribed in black, retaining differentiation through font size or the use of uppercase letters to emphasize more important geographic features. This visual hierarchy enabled chart users to quickly identify key locations and their significance (Fig. 3.4).

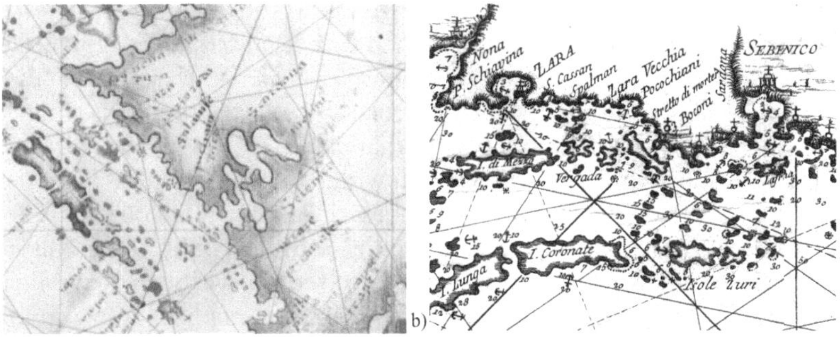

a) b)

Fig. 3.4 The hierarchy of toponyms: coloured in red on a manuscript chart: (**a**) Diogo Homem, 1570, and use of capitalization on a printed chart; (**b**) Lodovico Furlanetto, 1784 (Sources: (a) Hrvatski državni arhiv, Kartografska zbirka, Zagreb, HR-HDA-902, D.XIV.6; (b) Državni arhiv u Zadru, Kartografska zbirka, Zadar, HR-DAZD-383 No. 3.1)

5.2 *Renaming History: Toponyms in Political and Cultural Transformations*

Toponyms often carry a symbolic component, reflecting the motivations of those who selected them and the cultural context of their choice (Orth 1987). Ideologically motivated toponyms are particularly susceptible to change during periods of radical socio-political upheaval, as renaming shapes new identities and alters or erases symbolic pasts. Thus, toponyms become a reflection not only of places but also of cultures (Šakaja 2015). Within this study, the phenomenon of multiple naming (polionymy) is particularly significant, where a single geographic feature bears several names. This allows for a diachronic perspective on the chronology of naming and changes, as well as a synchronic analysis of cartographers' or patrons' attitudes towards a space (Skračić 2009; Crljenko and Faričić 2022). A notable example of polyonymy is the Adriatic Sea, whose nomenclature reflects diverse political and cultural influences. Names such as *Mare Adriaticum* and *Golfo di Venezia* testify to historical and political processes that have shaped perceptions of this space. The Latinized name *Mare Adriaticum* highlights cultural continuity from Greco-Roman times (Kozličić 1990; Faričić et al. 2023b), while *Golfo di Venezia* represents Venetian political ambitions and hegemonic propaganda. During the late medieval and early modern periods, Venice viewed the Adriatic as its economic and maritime domain, forming the basis of its economic and political dominance (Lane 1973; Vivo 2003). Out of 84 analysed charts, the Adriatic Sea was unnamed on 34, referred to as the Gulf of Venice on 29 and as the Adriatic Sea on 11, and named with both terms on eight, exclusively on printed charts. The reproduction of Venetian chart templates in European cartographic centres, particularly in the Netherlands, France, and England, facilitated the spread of that name. However, this reproduction was often technical and uncritical, without deliberate transmission of political messages. The name "Gulf of Venice" gradually fell out of use and was replaced by "Adriatic Sea," reflecting shifting political and cultural dynamics, particularly after Napoleon dissolved the Venetian Republic in 1797. The Venetian choice of *Golfo di Venezia* exemplifies the "foundational technology" (Giraut and Houssay-Holzschuch 2016) through which Venetian cartographers sought to legitimize a new political and cultural order on both regional and global scales, as the name became embedded in European cartographic traditions. This practice highlights the power of charts as instruments of political propaganda. Changes in the

Adriatic's nomenclature reflect the complexity of socio-political and geopolitical processes, making toponyms an important tool for understanding historical dynamics.

The analysis of the geographic name of Lokrum Island, situated in the vicinity of the Dubrovnik harbour, reveals the intricate layers of the communicative potential of charts and their role in shaping geopolitical discourse. As a strategic point and a target of hostile attacks, it became a focal point of conflict between the Republic of Dubrovnik and the Venetian Republic, culminating in the so-called Lokrum Crisis of the seventeenth century (Kunčević 2021). Venetian documentation from the 1620s often refers to the island as *Scoglio di San Marco* (Saint Mark's Island) (Makušev and Šufflay 1905). On late medieval and early modern charts, however, common names include *Lacroma* and *Croma*. The name *Scoglio di San Marco* first appears on a 1679 chart by Marccheto Fassoi. These variations in naming, ranging from the Venetian name to the local Lokrum (Fig. 3.5), illustrate how cartography records and reinterprets historical conflicts and cultural layers of space. The continued use of *Scoglio di San Marco* on charts following the fall of the Venetian Republic, including British and French charts and sailing directions of the nineteenth century,

a) b)

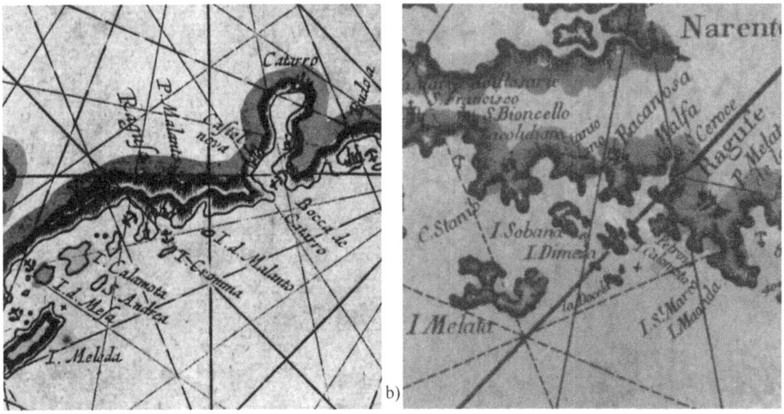

Fig. 3.5 Depictions of Lokrum island on early modern nautical charts of the Adriatic: (**a**) Lokrum named as *I. Cromma*; Pieter Goos, 1650; (**b**) Lokrum named as I. *St. Marco*; Joseph Roux, 1764 (Sources: (a) Private Collection Marco Asta, Bologna; (b) Nacionalna i sveučilišna knjižnica u Zagrebu, Zbirka zemljovida i atlasa, Zagreb, S-JZ-XVIII-116)

demonstrates the resilience of cartographic elements (Sutlović 2024a). While the name on Venetian charts served as a vehicle for disseminating geopolitical narratives, later European cartographers reproduced it without its original symbolic connotations. By reflecting and perpetuating the cultural and political ideologies of their creators, charts transcend their technical function and become tools of communication that shape historical and cultural understanding of space.

5.3 Island Naming Practices: Balancing Scale and Maritime Priorities

The naming of islands on Adriatic nautical charts during the Middle Ages and early modern period depended significantly on their socio-economic and navigational importance and the available space on the chart. In the absence of standardized generalization methods, the depiction of islands was subjective, relying on the amount of available information and the qualifications of the cartographer. The charts analysed indicate that relatively few islands were named, with no significant variation in the number of named near versus offshore islands, despite the higher number of near islands. Nonetheless, small islets like Galijula, Jabuka, Palagruža, and Pianosa were consistently depicted and named. For example, on charts at a scale of 1:1,000,000, an island with an area of 1 km^2 should be represented as 1 mm × 1 mm. Yet, small islets with areas below 1 km^2 were exaggerated and also named. These islets were depicted and named because of their critical role in terrestrial navigation. Failing to name or depict them would compromise maritime safety. Jabuka and Palagruža were unnamed on only 4% of the analysed charts, primarily those representing areas broader than the Adriatic Sea. The naming of small islands and islets reflects their vital role in maritime safety and efficient use of the Adriatic's maritime space. Even on charts at smaller scales, navigation-critical islets were regularly depicted and named, whereas larger islands with less navigational significance often remained unnamed. This practice highlights how cartographers prioritized maritime safety and navigational needs over strict adherence to technical scaling requirements. Through the consistent naming of key islets, charts served as communicative tools, enabling safe navigation and orientation in the Adriatic Basin while demonstrating cartographers' deep understanding of the specific needs of maritime traffic.

6 Cartographic Signs: The Language of Navigation

Cartographic signs are an essential component of any chart, as they enable effective communication between cartographers and users. Without them, the utilitarian value of charts would be almost non-existent. Their historical development reflects increasing complexity and sophistication, spurred by technological advancements in the early modern period (especially copperplate printing) (Robinson 1975; Verner 1975). However, during this time, there was no formal organization to guide cartographers in standardizing signs. In the Renaissance cultural ethos, which celebrated individual creativity, such constraints would have conflicted with humanistic values (Delano-Smith 2007). While semiotic inconsistency was common, particularly on topographic charts, maritime charts gradually moved towards a form of proto-standardization, adopting "customary signs" that facilitated clear and consistent communication. Safe navigation depended on unambiguous markings for natural hazards.

On the oldest surviving nautical charts, signs such as crosses or plus signs marked rocks/rocks awash, while red dots indicated sandy shoals. These were the only signs providing geomorphological information relevant to navigation (Campbell 1987). On the *Carte Pisane* (ca. 1270), only one sign for rocks/rocks awash is present. Over time, this sign gained several variations. Though arbitrary in origin, these signs were critical for safe navigation, denoting invisible or hard-to-spot dangers. While not all hazards were depicted, those that were significantly enhanced navigational safety. By 1325, portolan charts depicting the Mediterranean Sea displayed up to four different forms of rocks/rocks awash signs (Clawson 1979). The first chart of the Adriatic Sea to show multiple variations was Grazioso Benincasa's 1472 chart, which used three distinct signs for rocks/rocks awash. However, most Adriatic charts featured only two forms of this sign, typically a cross (+) and a cross with dots (⁜), with the former being more prevalent (Fig. 3.6). This distinction may reflect the categorization of "hazardous rocks" into visible, above-water rocks and submerged, less visible rocks awash. This sign with its variations appears consistently across the studied charts (Sutlović 2024b).

Sign for shoals first appeared on Vesconte's 1318 chart, depicted as red dotted lines or stippling, though the colour was later changed to black. This sign remained relatively consistent throughout the studied period. Clawson (1979) also identified a similar sign denoting "dangerous curves,"

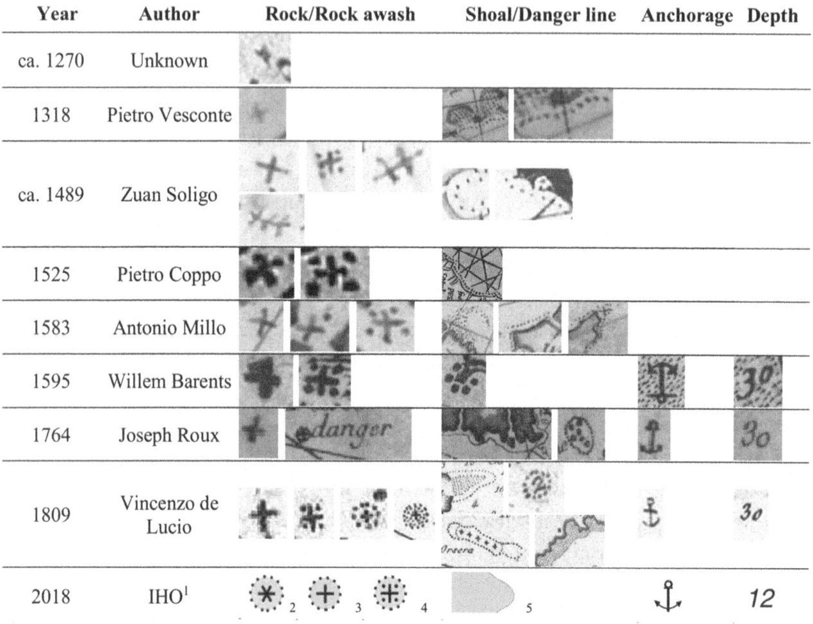

[1] International Hydrographic Organization (2021). B-410: 5, B-420:1-2, B-430: 1.

[2] Isolated rocks which cover and uncover (IHO, 2021, B-420:1),

[3] Rocks which are always underwater (IHO, 2021, B-420:2).

[4] Rocks which are awash at Chart Datum (IHO, 2021, B-420:2).

[5] A danger line (IHO, 2021, B-420:1).

Fig. 3.6 The exemplary sample of cartographic signs on nautical charts of the Adriatic Sea. Sources: Unknown author, ca. 1270, (Bibliothèque nationale de France, Paris, GE B-1118 (RES)); Pietro Vesconte, 1318, (Österreichische Nationalbibliothek, Vienna, Cod. 594 (Cimel. 20), 10v-11r); 9. Zuan Soligo, ca. 1489 (British Library, London, Egerton MS 73); Pietro Coppo, 1525 (Pomorski muzej Sergej Mašera, Piran, No. 1002307, T. 15); Antonio Millo, 1583 (Biblioteka Narodnowa, Warsaw, BN ZZK 0.2399); Willem Barents, 1595 (Stanford University Libraries, The Barry Lawrence Ruderman Map Collection, Stanford, Tooley, A-D:85, 268; K-P:10); Joseph Roux, 1764 (Nacionalna i sveučilišna knjižnica u Zagrebu, Zbirka zemljovida i atlasa, Zagreb, S-JZ-XVIII-116); Vincenzo de Lucio, 1809 (Sveučilište u Zadru, Znanstvena knjižnica, Zadar, 15188 D-20) marked by black dotted lines, indicating areas with reefs or rocks upon which waves broke. It is likely a sign denoting a danger line. Regardless of its exact interpretation, one thing is certain—caution is required when navigating the marked area. At times, distinguishing between islands, shoals, or zones dangerous for navigation was challenging, especially for navigators unfamiliar with the region. The sign for a danger line first appeared on Abraham Cresque's chart (ca. 1370—1380) south of Corfu. While abstract, the dotted texture evoked the appearance of shallow, sandy seabed, maintaining its design with minimal changes over time. This sign, albeit modified, continues to be used on modern nautical charts.

The next sign to emerge is the sign for safe anchorage, a stylized depiction of an anchor. It is a representational or pictorial sign first used by Pîrî Reis (1520), though not on Adriatic Sea charts. It became a standard feature of nautical charts by the late sixteenth century, with William Barents (1595) introducing it for Adriatic anchorages. The sign remained unchanged throughout the studied period, undergoing only minor stylistic modifications. Barents also introduced numerical depth markings in selected Adriatic harbours, marking another introduction in cartographic signs on charts depicting the Adriatic. Following Barents, Tentivo marked soundings in the Adriatic at the end of the seventeenth century, succeeded by Ioannes van Keulen in 1700. By the second half of the eighteenth century, soundings were regularly included on charts, even though the first hydrographic survey of part of the Adriatic was conducted from 1806 to 1809, by Charles-François Beautemps Beaupré (Horvat et al. 1999; Kozličić 2006). Along with charts of the Adriatic, Jacques-Nicolas Bellin (1771) and Roux (1779) created detailed harbour charts featuring soundings. The soundings were of great importance for sailors, especially when navigating shallow areas.

Additional signs emerged during the early modern period. Buoys appeared on charts from the late sixteenth century, followed by signs for lighthouses and other light signals in the second half of the seventeenth century, and currents by the early eighteenth century. On Adriatic Sea charts, however, these signs did not become common until the late eighteenth or early nineteenth century. V. De Lucio (ca. 1792–1796) first used the sign for sea current, an arrow with a feathered tail. Giovanni Grubas's 1803 chart included both current signs and buoy markers, marking the beginning of their consistent use in Adriatic cartography (Sutlović 2024b). Labels for seabed types were not introduced until 1820, despite the British having developed a system of conventional signs and abbreviations for seabed types as early as 1776.

The transition from the eighteenth to the nineteenth century was pivotal for the standardization of cartographic signs. This period saw the institutionalization of nautical cartography with the establishment of hydrographic offices and the application of hydrographic surveys, laying the groundwork for modern nautical charts. The continuity of rudimentary signs for navigational safety (in use since the earliest nautical charts) and anchorages (introduced in the late sixteenth century) underscores the enduring utility of these features.

The communicative potential of cartographic signs grew with the introduction and standardization of new signs, enhancing the clarity of nautical charts regardless of their origin or the region depicted. This development established a cartographic system that turned nautical charts into a universal "language" for navigators, facilitating the exchange of vital navigational information across cultural and geographic boundaries.

7 The Complexity of Map Scales

A map scale is a fundamental mathematical element of any map/chart, contributing to the decoding of its content, particularly the linear and areal dimensions as well as spatial relationships. While users of old nautical charts may not have fully understood the concept of scale as "the ratio of distances on a map to corresponding distances on the surface of an ellipsoid or sphere" (Frančula et al. 2020: 122), they were familiar with its basic function. On the studied charts, a graphical scale was regularly included. It was usually shown as a horizontal line or several lines with equal-distance divisions that typically represented a number of miles or leagues within a one-degree arc along the equator, shrunk size-wise in accordance with the map scale.

The communicative potential of a nautical chart is significantly increased when the graphical scale was expressed in multiple units of length. Until the nineteenth century, various states, regions, and even cities used their own units of distance. A graphical scale can be converted into numerical form and used for calculations, but this requires knowledge of the unit's length and the measurement system applied by the cartographer (Mušnjak 1982). This process seems straightforward, but research has shown that cartographers often misnamed individual units of distance.

Of the 84 analysed nautical charts of the Adriatic Sea, 42 included a graphical scale without any unit of distance, a feature typical for portolan charts. Adolf E. Nordenskiöld (1897) coined the term "portolan mile" for this unit. Numerous debates have arisen about the length of this mile, resulting in various values, with the most recent calculations estimating its length at 1.25 km (Marelić 2025). In the analysed corpus of charts, the first author to include a unit of distance alongside the graphical scale was Alonso de Santa Cruz (circa 1545).

By the late sixteenth century, and particularly from the mid-seventeenth century onwards, cartographers began to specify not only the unit of distance but also the number of these units "within a degree," or within the arc corresponding to one degree of a great circle. W. Barents (1595) was the first to note how many of each unit fit within one degree. Barents identified three units of distance and provided their equivalents: *Miliaria Italica/Italiaenische milen* (70 per degree), *Hispanicae Leuce/Spaensche milen* (17.5 per degree), and *Miliaria Germanica/Duytsche milen* (15 per degree) (Fig. 3.7).

Excluding *Carta di Cabotaggio del Mare Adriatico* published by the Military Geographic Institute in Milan (1822–1824), which was the result of hydrographic surveying, as many as 16 charts in the analysed corpus featured three or more units of distance alongside their graphical scales.

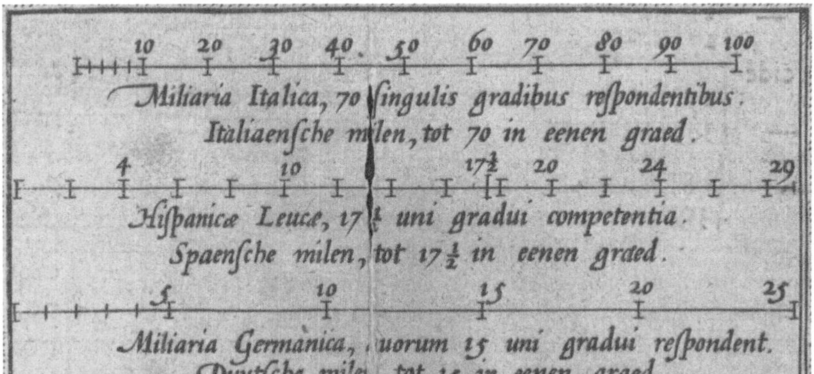

Fig. 3.7 Scales and units of distance on Barents's nautical chart of the Adriatic Sea, 1595 (Source: Stanford University Libraries, The Barry Lawrence Ruderman Map Collection, Stanford, Tooley, A-D:85, 268; K-P:10)

The most commonly noted units were: Italian mile (60 or 75 per degree), Spanish league (17.5 per degree), German or Dutch mile (15 per degree), and English or French league (20 per degree).

Due to uncritical copying of content, errors related to scale were also made by authors of nautical charts. The most frequent error involved confusing the terms leagues and miles. The modern league is generally considered to be about three miles, either statute or nautical (Moody 1950). Historically, however, the length of a league has varied significantly, ranging from approximately 2.4 to 4.6 miles depending on regional and temporal contexts (Chardon 1980). As early as 1595, W. Barents mistranslated *Hispanicae Leuce* as *Spaensche milen*. Analysis has shown that this unit referred to leagues, not miles. Similar terminological errors occurred among cartographers for Italian, German, and French units, indicating systematic issues in cartographic practices. The lack of sources in the literature explaining these errors leaves it unclear whether they arose from a single perpetuated mistake or another cause, though the first explanation is likely. The inconsistency in the use of the terms "mile" and "league" may have created problems for sailors, reducing the communicative potential of nautical charts.

The nautical charts with the greatest number of distance-units on their graphical scales were created by Keulen (1700), François Berthelot (1700), and Roux (1764). These Adriatic Sea charts feature four graphical scales with different units. While all correctly noted the quantitative value of each unit within the arc corresponding to one degree of a great circle, some included incorrect terminological designations of the units. Over time, the communicative potential of charts regarding scales improved. The chart with the highest communicative potential in this respect is *Carta di Cabotaggio* (1822–1824). In addition to scales for converting units of distance such as *Piedi di Parigi*, *passi Veneti*, *passi Napoletani*, and *Fadom Inglesi*, this chart includes numerical scales (e.g. 1:175.000, 1:70.000, 1:35.000, 1:25.000) and units such as *Metri*, *Tese di Francia*, *miglia geografici*, and *chilometri geografici*.

Based on the analysed corpus of nautical charts, it can be concluded that cartographers struggled with the terminological distinction between "mile" and "league," often using these terms interchangeably. The problems with unit terminology illustrate how semantic confusion can affect the interpretation and use of cartographic information. These challenges highlight the critical need for standardization and clarity in cartographic practices to enhance navigational safety and the utility of charts.

8 Communication Noise: Challenges of Island Naming and Positioning

In the context of communication, noise refers to any interference that negatively affects the reception of signals within a communication system. While not every disturbance obstructs the decoding of information, it can lead to difficulties in understanding messages (Shannon and Weaver 1949).

An analysis of toponymic data has revealed an issue with naming the island of Ist. Examining the chain of offshore islands from northwest to southeast between Ilovik and Dugi Otok, it is clear that the islands appear in the following order: Premuda, Škarda, Ist, and Molat. Among these, cartographers consistently named Premuda, Škarda, and Molat, while Ist was often unnamed or misnamed. On the oldest studied chart, the *Carte Pisane* from approximately 1270, Molat (*Melada*) is the only named island in this chain. On the chart by P. Vesconte (1318), Premuda (*premuda*), Škarda (*sechera*), Ist (*est*), and Molat (*mellada*) are all named. However, by the next studied chart, Giacomo Giroldi's chart from 1426, the issue with geographic names begins to emerge. Giroldi named Molat (*mellada*) and another nearby island northwest of Molat (*zanpontello*). This error persisted throughout the early modern period. The first cartographer after Vesconte to correctly name the island of Ist on a nautical chart was De Lucio (ca. 1792–1796), who added the letter "L" to the name, rendering it as *LISTO* (Fig. 3.8). By the early nineteenth century, this error disappeared, and Ist was regularly labelled with its proper toponym. Zapuntel is the name for a settlement on Molat near the strait separating Molat and Ist. It is presumed that this error originated from uncritical copying and a lack of geographical knowledge, leading to Ist being mislabelled as Zapuntel.

Other recurring errors were also identified, such as the incorrect positioning of the island of Premuda. Some cartographers erroneously placed Premuda between Susak and Ilovik, although in reality, it lies southeast of Ilovik. This mistake was made by W. Barents (1595), Willem J. Blaeu (1621), Pieter Goos (1650), Hendrick Doncker (1655), Francesco M. Levanto (1664), John Seller (1677), Pierre Van der Aa (1700), and John Mount and Thomas Page (1750), leaving the error uncorrected for over 250 years. In sailing directions by G. Benincasa, W. Barents, W. J. Blaeu, and J. Seller, there are no detailed descriptions of Premuda, but illustrations of the island are accompanied by brief comments. For instance, Barents, Blaeu, and Seller note beneath their illustrations of

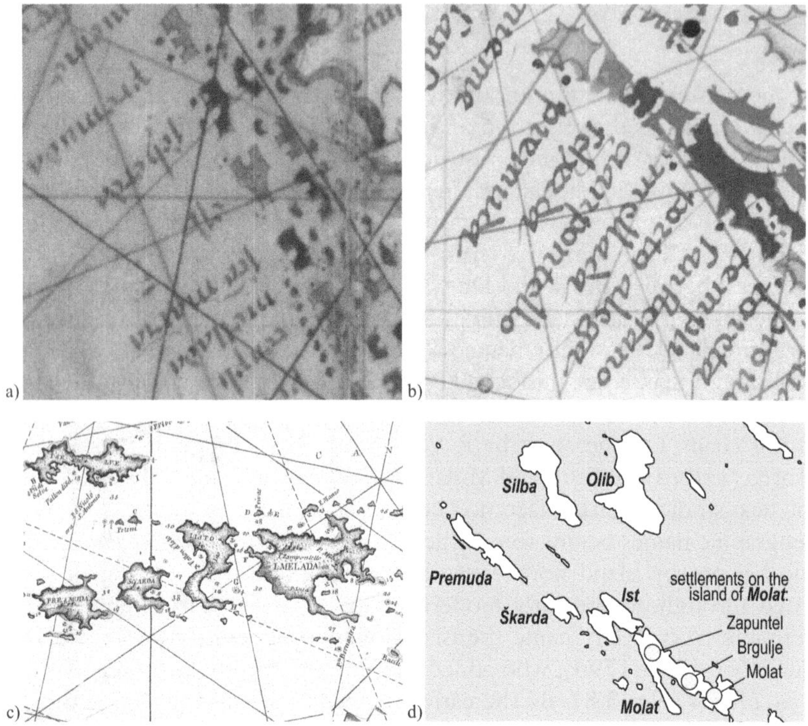

Fig. 3.8 Snippets from charts on which Premuda, Škarda, Ist, and Molat are represented: (**a**) P. Vesconte, 1318; (**b**) G. Giroldi, 1426; (**c**) V. de Lucio ca. 1792–1796; (**d**) Nautical chart Silba—Pag 100–19, 2003 (Sources: (a) Österreichische Nationalbibliothek, Vienna, Cod. 594 (Cimel. 20), 10v-11r; (b) Biblioteca Nazionale Marciana, Venice, It. VI, 212 (=5694); (c) Hrvatski državni arhiv, Kartografska zbirka, Zagreb, T 3 B + 5 D—HR-HDA-902 E.IV.13-a; (d) marineregions.org, obtained on April 23, 2025)

Premuda that the island lies between Lošinj and Škarda, even though their charts misplace its location.

These are just examples of errors that can be found on the studied charts which underscore the importance of a detailed understanding of geographic spaces and a critical approach to the use of nautical charts. The creation of accurate and detailed charts was essential for navigational safety and the efficient communication within cartographic systems.

Communication noises could have had practical consequences for navigation. Decoding messages from charts required not only familiarity with signs and other conventional elements of cartography but also awareness of potential noise within the communication system.

9 Conclusion

Early modern nautical charts of the Adriatic Sea were not merely technical tools for navigation but multifaceted media for conveying geographical, hydrographic, navigational, political, economic, cultural, and confessional information. Their analysis reveals the depth and complexity of the messages embedded within their content, including imagological elements, toponyms, cartographic signs, scales, units of distance, and communication noise. Communication theory provides a lens through which these charts can be viewed as intricate channels for transmitting information between cartographers and users. Each element of the communication model is present in the creation and interpretation of nautical charts, where cartographers encoded geographic realities into visual representations designed to be decoded by users. This process required the careful selection of geographic and other information to convey layered messages, while users needed specific knowledge and skills to accurately interpret that information.

Through precise depictions of coastlines, navigational routes, and strategic points, these charts facilitated the effective communication of vital information critical for maritime traffic. Consequently, nautical charts directly influenced socio-economic development, as maritime activities often served as either a driver or catalyst of development. Late medieval and early modern nautical charts extended beyond basic navigational information to include vexillological, heraldic, and allegorical depictions, such as the winged lion of St Mark symbolizing the Venetian Republic, the Habsburg coat of arms, or personifications of Venice. These elements illustrate the communicative potential of charts in defining territorial belonging, confessional dominance, and political relationships.

The analysis of toponyms on Adriatic Sea charts demonstrates how navigational needs, cultural influences, and political dominance, particularly by the Venetian Republic, shaped cartographic practices. Hydronyms like *Mare Adriaticum*, reflecting the cultural continuity of Latin heritage, and *Golfo di Venezia*, emphasizing Venetian political aspirations, underscore how cultural and political influences shaped spatial perceptions.

Varied names for the same objects, such as *Lacroma* and *Scoglio di San Marco* for Lokrum, point to conflicts in geopolitical discourse. The hierarchical representation of toponyms, employing colours, font sizes, and styles, alongside the consistent naming of small islands as key navigational markers, underscores the methodical approach cartographers took to highlight significant locations for safe navigation.

The development and standardization of cartographic signs for rocks/ rocks awash, shoals, anchorages, and other features reflect the continuous efforts to enhance nautical charts for improved navigational safety. Comparative analyses confirm that these signs developed into standardized elements, enabling the reading of critical information about navigational risks. The persistence of rudimentary cartographic signs for navigation safety, from the earliest nautical charts to the present day, demonstrates a remarkable continuity in nautical cartography. Hydrographic surveys and the institutionalization of cartography provided detailed, universally comprehensible representations of significant features and phenomena.

By the late sixteenth century, cartographers began incorporating multiple units of distance on the same chart, a practice pioneered for the Adriatic region by W. Barents in 1595. This introduction increased the communicative potential of charts, enabling comparisons of units used across different regions. Common units included Italian, German, and Dutch miles, as well as Spanish, French, and English leagues. The communicative potential of scales culminated in the early nineteenth century with *Carta di Cabottagio del Mare Adriatico*, marking a high point in the representation of measurement systems.

Errors in the charting of islands such as Ist and Premuda illustrate instances of communication noise that undermined the accuracy and reliability of navigational information. Often resulting from limited geographic knowledge or uncritical copying of older sources, these errors posed challenges to effective communication and safe navigation. The correction of such mistakes highlights the development of cartographic practices aimed at improving the reliability of charts as navigational tools, an effort made possible by the institutionalization of cartography and its reliance on systematic surveys.

The development of nautical charts of the Adriatic, most of which were of Venetian origin, aligns with broader European trends in cartography, while also reflecting regional characteristics, and the unique imprint of Adriatic maritime and geopolitical conditions. Despite this alignment, there was a noticeable lag in the adoption of certain cartographic signs compared to Dutch, British, and French cartographers during the early modern period. This lag, evident in the late stages of the early modern period, highlights the decline of Venice as a leader in cartographic originality and technological innovation, as other European centres surpassed it.

Early modern nautical charts of the Adriatic Sea are not only documents for exploring the historical development of cartographic skill but also vital witnesses to the cultural, political, and social dynamics of their time. Their detailed analyses reveal complex layers of meaning that transcend the technical aspects of cartography, affirming that charts are indeed communication media with a significant impact on historical and, to some extent, contemporary perceptions of the Adriatic. This study highlights the importance of nautical charts as dynamic communication media, enriching the understanding of nautical cartography's historical development and framing these charts not only as mirrors of social and economic processes in the Adriatic but also as tools that shaped them. Early modern cartographers did more than depict; they conceptualized and shaped the Adriatic's coastal, insular, and maritime space as a zone of multifaceted contact and conflict.

References

Barnes T, Gregory D (1997) Glossary. In: Barnes T, Gregory D (eds) Reading human geography: the poetics and politics of inquiry. Arnold, pp 503–505

Bocharov MK (1967) Teoreticheskie problemy kartografii i avtomatizatsiya kartograficheskovo proizvodstva. Izvestiya vysshikh uchebnykh zavedeny. Geodeziya i aerofotosyomka 4:19–26

Braudel F (1997) Sredozemlje i sredozemni svijet u doba Filipa II. Izdanja Antibarbarus

Brown LA (1979) The story of maps. Dover Publications

Campbell T (1987) Portolan charts from the Late Thirteenth Century to 1500. In: Harley J, Woodward D (eds) The history of cartography, volume 1, cartography prehistoric, ancient and medieval Europe and the Mediterranean. University of Chicago Press, Chicago, pp 371–463

Chardon R (1980) The linear league in North America. Ann Assoc Am Geographers 70(2):129–153. https://doi.org/10.1111/j.1467-8306.1980.tb01304.x

Clawson MG (1979) The evolution of symbols on nautical charts prior to 1800 [Master Thesis]. University of Maryland

Crljenko I, Faričić J (2022) Toponymic twins, polyonymy in Croatia. In: Loth C-R (ed) Standardisation and the wealth of place names—aspects of a delicate relationship. Sun Media Bloemfontein, pp 167–189

Culcasi K (2008) Cartographic constructions of the Middle East. Doctoral thesis. Syracuse University

Davenport Adams WH (1869) The queen of the Adriatic; or, Venice past and present. Thomas Nelson & Sons

de Vivo F (2003) Historical justifications of Venetian Power in the Adriatic. Journal of History of Ideas 64(2):159–176. https://doi.org/10.2307/3654123

Delano-Smith C (2007) Signs on Printed Topographical Maps, ca. 1470-ca. 1640. In: Woodward D (ed) The history of cartography: cartography in the European Renaissance. The University of Chicago Press, pp 528–590

Eckert M (1908) On the nature of maps and map logic. Bull Am Geographical Soc 40(6):344–351. https://doi.org/10.2307/198027

Faričić J (2011) Geografski aspekti proučavanja toponima. In: Skračić V, Faričić J (eds) Geografska imena, Zbornik radova. Sveučilište u Zadru, HGD, pp 51–66

Faričić J, Kljajić I, Mirošević L, Mlinarić D (2023a) Symbolism of compass roses on early modern nautical charts of the Adriatic Sea. KN – J Cartography Geographic Inf 73:19–37. https://doi.org/10.1007/s42489-022-00129-z

Faričić J, Selva O, Umek D (2023b) Geographical names of the Adriatic Sea on medieval and early-modern maps and nautical charts. J Hist Geography 82:68–80. https://doi.org/10.1016/j.jhg.2023.09.007

Frančula N, Lapaine M, Jazbec I-P (2020) Mjerilo. In Kartografski rječnik. Hrvatsko kartografsko društvo i Naklada Dominović, p 122

Garret M (2001) Venice: a cultural and literary companion. Signal

Giraut F, Houssay-Holzschuch M (2016) Place naming as Dispositif: toward a theoretical framework. Geopolitics 21(1):1–21. https://doi.org/10.1080/14650045.2015.1134493

Griffith E (2005) The Winged Lion of St. Mark: Logo of Venice, Inc. European Business History Association 2005 Conference

Harley JB (1988) Maps, knowledge, and power. In: Cosgrove D, Daniels S (eds) The iconography of landscape: Essays on the symbolic representation, design and use of past environments. Cambridge University Press, pp 277–312

Harley JB (1992) Deconstructing the map. In: Barnes T, Duncan J (eds) Writing worlds. Discourse, text and metaphor in the representation landscape. Routledge, pp 231–247

Horodowich E (2021) Why were there no Continental Allegories in Renaissance Venice? The Amerasian Personifications of Giuseppe Rosaccio. In: Horowitz MC, Arizzoli L (eds) Bodies and maps early modern personifications of the continents. Brill, pp 217–237

Horvat S, Duplančić T, Željeznjak Ž (1999) Thematic nautical charts and handbooks for the sea area of the Eastern Adriatic Coast. Geodetski list 53(2):151–156

International Hydrographic Organization (2021) Regulations of the IHO for international (INT) Charts and Chart Specifications of the IHO. International Hydrographic Organization

Keates JS (1984) The cartographic art. Cartographica 21(1):37–43. https://doi.org/10.3138/F27N-7125-8Q25-8670
Kitchin R, Dodge M (2007) Rethinking maps. Progr Hum Geography 31(3):331–344. https://doi.org/10.1177/0309132507077082
Koláčný A (1969) Cartographic information—a fundamental concept and term in modern cartography. Cartographic J 6(1):47–49. https://doi.org/10.1179/caj.1969.6.1.47
Koláčný A (1977) Cartographic information—a fundamental concept and term in modern cartography. Cartographica 14(1):39–45. https://doi.org/10.3138/N587-4H37-2875-L16J
Kozličić M (1990) Historijska geografija istočnog jadrana u starom vijeku. Književni krug
Kozličić M (2006) Istočni Jadran u djelu Beautemps-Beaupréa. Hrvatski hidrografski institut
Kunčević L (2021) O „drskosti" Dubrovčana: mletačko-dubrovački sukob u ljeto 1630. godine. Anali Zavoda za povijesne znanosti Hrvatske akademije znanosti i umjetnosti u Dubrovniku 59:81–114. https://doi.org/10.21857/mnlqgc5k2y
Lane FC (1973) Venice: a maritime republic. John Hopkins University Press
Lapaine M, Midtbø T, Gartner G, Bandrova T, Wang T, Shen J (2021) Definition of the map. Adv Cartography GIScí Int Cartographic Assoc 3(9):1–6. https://doi.org/10.5194/ica-adv-3-9-2021
MacEachren AM (2004) How maps work: representation, visualization, and design. The Guilford Press
Makušev VV, Šufflay M (1905) Isprave za odnošaj Dubrovnika prema Veneciji. Starine 31, JAZU
Maletzke G (1998) Kommunikationswissenschaft im Überblick: Grundlagen, Probleme, Perspektiven. VS Verlag für Sozialwissenschaften Wiesbaden
Marelić T (2025) Are Portolan Charts and Portolan Mile Geometrically Rooted in Classical Antiquity? A Cartometric Analysis of al-Shirazi's "Greek Map" and the Pisane, Lucca, Avignon, and Cortona Charts. Cartographica 59(4):143–160. https://doi.org/10.3138/cart-2024-0029
Marković M (1993) Descriptio Croatiae: Hrvatske zemlje na geografskim kartama od najstarijih vremena do pojave prvih topografskih karata. Naprijed
Miller W (1903) The Ionian Islands under Venetian Rule. English Hist Rev 18(70):209–239. https://doi.org/10.1093/ehr/XVIII.LXX.209
Mlinarić D, Gregurović S (2011) Kartografska vizualizacija i slika Drugoga na primjeru višestruko graničnih prostora. Migracijske i Etničke Teme 27(3):345–373
Monmonier M (2004) Rhumb lines and map wars: a social history of the mercator projection. The University of Chicago Press
Moody AB (1950) The nautical mile. Int Hydrographic Rev 27(1):77–85

Mountz A (2009) The other. In: Gallaher C, Dahlman C, Gilmartin M, Mountz A, Shirlow P (eds) Key concepts in political geography. Sage, pp 328–338
Mušnjak T (1982) Mjerila na starim zemljopisnim kartama. Arhivski vjesnik 25(1):47–64
Nordenskiöld AE (1897) Periplus: an essay on the early history of charts and sailing directions. P.A. Norstedt & Söner
Orth DJ (1987) Principles, policies, and procedures: domestic geographic names. U.S. Board on Geographical Names
Pápay G (2018) Max Eckert and the foundations of modern cartographic praxis. In Kent AJ, Vujaković P (eds) The Routledge handbook of mapping and cartography. Routledge, pp 9–28
Pedani MP (2006) Il Leone di San Marco o San Marco in forma di Leone?. Archivio Veneto, Serie V, vol CLXVI: 185–190
Pickles J (2004) A history of spaces. Cartographic reason, mapping and the geo-coded world. Routledge
Rizzi A (2012) I leoni di San Marco. Cierre edizioni
Robinson A (1952) The look of maps: an examination of cartographic design. The University of Wisconsin Press
Robinson AH (1975) Mapmaking and map printing: the evolution of a working relationship. In: Woodward D (ed) Five centuries of map making. The University of Chicago Press, pp 1–25
Robinson AH, Petchenik B (1975) The map as a communication system. Cartographic J 12(1):7–15. https://doi.org/10.1179/caj.1975.12.1.7
Šakaja L (2015) Uvod u kulturnu geografiju. Leykam interbational d.o.o
Salitchev KA (1970) The subject and method of cartography: contemporary views. Cartographica Int J Geographic Inf Geovisual 7(2):77–87. https://doi.org/10.3138/92G4-N258-7879-1796
Shannon SE, Weaver W (1949) The mathematical theory of communication. Urbana
Siniscalchi S, Palagiano C (2018) The place names of the Middle East Before and After Ptolomeic Cartography: an emblematic selection from ancient maps. Cartographic J 55(3):205–216. https://doi.org/10.1080/00087041.2017.1413788
Skok P (1950) Slavenstvo i romanstvo na jadranskim otocima. Jadranski institut JAZU
Skračić V (2009) Višeimenost u jadranskoj nesonimiji (na primjeru Kornatskog otočja). Geoadria 14(1):141–163. https://doi.org/10.15291/geoadria.560
Sutlović J (2024a) Conflict of geopolitical discourse illustrated by the example of the geographical name of Lokrum island. Geoadria 29(1):54–91. https://doi.org/10.15291/geoadria.4451

Sutlović J (2024b) Signs related to the safety of navigation on late medieval and early modern nautical charts of the Adriatic Sea, circa 1270–1824. J Hist Geography 89:27–45. https://doi.org/10.1016/j.jhg.2024.11.006

Tubbs S (2003) Human communication: principles and context (9th ed). McGraw-Hill

Verner C (1975) Copperplate printing. In: Woodward D (ed) Five centuries of map making. The University of Chicago Press, pp 1–25

Wood D (1992) The power of maps. Guilford Press

Wood D, Fels J (1986) Design on signs/myth and meaning in maps. Cartographica 23(3):54–103. https://doi.org/10.3138/R831-50R3-7247-2124

Zorzi A (1980) Una città una Repubblica un impero. Venezia (697–1797). Mondadori

CHAPTER 4

Navigare et Regere: Venetian Cartography Through the Centuries—From Planning Tool to Symbol of Power

Orietta Selva and Dragan Umek

1 Introduction

According to Franco Farinelli (1992), geographic representation can be seen as a *system of peculiar signs*, and thus its understanding can be defined as decoding and in this regard maps become almost a "cyphered document"; hence, its investigation can focus either on the signified or on the significant. The first approach is probably the most common; the iconographic document is taken as a source of a plethora of information about the landscape and seascape. A different level of abstraction is, however, necessary for the second approach: disregarding the meaning of the signs, it focuses on the hidden purposes. This second aspect induces one to consider maps and nautical charts as documents able to offer cognitive visions capable of deciphering, both on the syntagmatic and paradigmatic level, the sign code of maps and nautical charts (Barsanti 1985: 55–58; Licini 2000: 25).

O. Selva (✉) • D. Umek
Department of Humanities, University of Trieste, Trieste, Italy
e-mail: orietta.selva@dsgs.units.it; dragan.umek@dsgs.units.it

J. Faričić, T. Marelić (eds.), *Early Modern Nautical Charts of the Adriatic Sea*, https://doi.org/10.1007/978-981-96-8046-7_4

Through the interpretation of the cartographic language, as well as the diverse semiotics used by cartographers, it is possible to collect extensive information, even if mediated by the author's subjective perspective. Cartography, as an expression of the dynamic interrelation of technological development and political interests, assumed different roles over time. These functions were, in the past as well as nowadays, determined by the cultural background, opportunity, specific needs, and scope; thus, from this perspective, cartography is an essential tool for the study of geopolitical dynamics (Boria 2020).

Regarding the cartographic depictions of the Mediterranean and, more specifically, of the Adriatic, extensive research has been conducted across different periods and using diverse methodologies. These range from documentary investigations and historical-cartographic studies rooted in positivist and evolutionist traditions (Marinelli 1881; Almagià 1929; Caraci 1959; Ferro 1992; Lago 1989; Quaini 1994; Astengo 2000; Valerio 2007) to more recent interpretations focusing on the symbolic or semiotic significance of cartographic documents. Examples include Harley's work on the "power of maps" (1989), Schultz's "maps as metaphors" (1990), and contributions by other scholars (Wood and Fels 1992; Casti Moreschi 1997, 1998; Heffernan 2007; Milanesi 2008; Sturani 2008; Farinelli 2003, 2009; Dodge et al. 2009; Edney 2019; Dai Prà 2010). Building on these studies, a distinction has emerged—albeit not universally accepted—between "technical" maps, created using scientific criteria, and "symbolic" maps, which function as visual metaphors for abstract models. This distinction has sparked debate over whether it is justified to consider all large-scale cartographic representations—whether wall maps or paper maps—displayed in public buildings exclusively, or at least primarily, as allegorical products and, therefore, only in terms of their metaphorical dimension (Schulz 1990).

In the case of Venice, maps and globes were a constant presence in the chambers of the Doge's Palace. From as early as the fourteenth century, the *Sala dello Scudo*—a representative hall where the Doge held audiences and received foreign guests—featured a series of cartographic depictions intended to showcase the vast territories of the Republic and, by extension, to celebrate its economic and political power. Over time, many of these depictions were lost due to the repeated fires that tragically marked the history of the palace, though they were subsequently replaced or supplemented with new works (Milanesi 2012).

For instance, after the 1483 fire destroyed the initial two maps depicting Italy and a world map by Antonio Leonardi, the same cartographer repainted a new map of Italy two years later, though the world map was not replaced until 1530. Historical sources reveal that during the sixteenth century, a series of maps created by renowned cartographers such as Giovan Battista Ramusio, Giovanni Domenico Zorzi, and Giacomo Gastaldi adorned the walls of the hall. In the eighteenth century, under the commission of Doge Marco Foscarini, all the maps were redone by Francesco Griselini. The Venetian cartographer and polymath followed the sixteenth-century model while adding new paintings depicting the voyages of famous Venetian explorers—Nicolò and Antonio Zen, Pietro Querini, Alvise da Mosto, and Giovanni Caboto—recounting the routes that had contributed to Venice's greatness (Pavanetto 2019: 53–58).

Marica Milanesi (2008: 191) also explored this topic, asserting that "the display of one's territorial portrait, in the form of views or cartography, represents one of the most traditional forms of power manifestation by a prince, common throughout Europe and already practiced in classical Rome. However, it is risky to attribute an exclusively symbolic and self-celebratory value to all [...] cartographic drawings present in palaces, whether public or private, between the Middle Ages and the Renaissance. It seems crucial to consider the differences between the types of territories represented, the locations of the various cycles, and the periods in which they were created. In the case of highly detailed depictions of small areas under actual political or economic control, maps – even large-scale mural maps displayed in palaces – serve to provide precise information about a territory and document its possession". Furthermore, these maps assume an official documentary value precisely because they are displayed in a public palace, the seat of the Doge's authority and the administration of power, thereby ensuring the authenticity and fidelity of the representation. In this sense, the case of the project assigned to Cristoforo Sorte illustrates the degree of caution Venice applied in the use of such maps. In 1578, the Veronese engineer and cartographer was commissioned to produce a large mural map of the entire Venetian mainland domain, intended for public display in the Senate Hall. However, the project was revoked out of concern over revealing sensitive information about Venetian territories. Instead, Sorte was instructed to create five detailed maps of the Venetian provinces to be stored secretly in a designated room within the palace (Cucagna 1964: 148).

2 Cartographies of Territories and Seas

The history of the Adriatic between the thirteenth and eighteenth centuries is, for the most part, the history of Venice: the queen of commercial trade and a preeminent maritime republic. The reasons for its power were undoubtedly tied to the initiatives and remarkable organizational skills of the Venetians, but also to the favourable geographical and cultural conditions that made the Adriatic a vital corridor linking Europe and the Eastern Mediterranean. This "liquid territory", as Fernand Braudel (1976) describes it, was a north–south axis, longer than it was wide, bordered to the north by the Alps, to the west by the low and marshy Italian coastline (and the Apennines in the hinterland), to the east by the numerous Quarner and Dalmatian islands (and Dinarides in hinterland) and low and sandy Albanian coast, and to the south by the narrowing of the Strait of Otranto. This narrowing was the defining feature of the basin, providing it with a sense of unity and uniqueness. Controlling this gateway (especially the island of Corfu) effectively meant dominating the entire Adriatic, thereby holding sway over what was considered one of the most important "gulfs" of the Mediterranean—a sea that was "a thousand things at once", a mosaic of civilizations, a succession of landscapes and seas whose waters were crossed by countless maritime routes weaving together its shores.

In this context, Venice assumed a unifying role, positioned as it was at the terminus of the "Great Sea" and serving as a true hub of exchanges between land and water. Its function was to act as a bridge between north and south, linking the routes to Central Europe with the routes to the Levant. This role was executed and upheld with absolute rigour in what the Serenissima referred to as *its Gulf*, a claim reinforced with determination in official documents that clearly articulated the concept of "dominion" as a natural right, justified by Venice's geographical position, rooted in the sea and the lagoon (Luzzatto 1995; Ivetić 2007).

The power of the Serenissima was not confined solely to commerce and trade; it extended across various domains, encompassing political, social, cultural, administrative, environmental, urban, military, and health-related spheres. This included the construction of fortifications and defensive works, the management of nature, borders, territories, lagoons, coastlines, rivers, forests, and much more. Over time, its primacy required the creation of a robust cartographic apparatus, remarkable for both its quantity and quality. These maps served as "representational models", capable of transposing onto paper the reality that Venetian interests were actively shaping (Selva 2016).

Navigare et regere—to navigate and to govern—are two key terms that encapsulate the fundamental purposes and values Venice consistently attributed to its cartographic production. It is noteworthy that the Republic of Venice was the first state to establish an administrative cartographic system; indeed, as early as the fifteenth century, maps were employed as tools of territorial policy, aimed at controlling and organizing the land (Casti Moreschi 1993).

This work seeks to provide a brief reflection on the evolution of Venetian cartography through a semiotic analysis of select historical documents produced between the fourteenth and eighteenth centuries. Particular attention will be given to the diverse characteristics these maps and nautical charts acquired in relation to cultural contexts, political circumstances, economic demands, interests tied to navigation, and the control and management of specific territorial areas, as well as the definition of new urban layouts.

Venetian cartographic production encompasses a wide array of iconographic types, ranging from *Isolari* to travel itineraries, nautical charts, chorographic maps, administrative documents, and micro-regional maps. Each type fulfilled one of the many functions of the state or catered to the private interests of nobles, merchants, and landowners. Thus, cartographic documents were used not only as tools for territorial governance but also as instruments to certify and celebrate the power and grandeur of the Republic of Venice. This is evidenced by inscriptions such as *Golfo di Venezia ovvero Mare Adriatico* ("Gulf of Venice or Adriatic Sea") on Venetian maps, which persisted long after the Republic's fall in 1797.

Attempting to conceptualize this diverse cartographic production into distinct typologies highlights the complexity of Venice's political and cultural landscape and the Serenissima's geopolitical vision of its dominions. To achieve this, we will draw on the framework proposed by Emanuela Casti (1999: 48), who categorizes Venetian cartography into two primary types: *Carte da navigare* (nautical charts), which depicted various coastal features, and *Carte da terra* or *Carte dello stato da terra e da mar* (maps of the land and maritime state), which were essential for the political and economic management of the Venetian state.

2.1 "*Carte da Navigare*"

Within the broad category of *Carte da navigare* (nautical charts), we can identify documents that served two distinct purposes, either exclusively or simultaneously. Some of these charts were primarily functional tools for navigators, while others were more descriptive. The latter aimed to disseminate geographic knowledge and construct an iconographic image of Venice's maritime and territorial dominance, celebrating its power (Biadene 1990).

The first category comprises charts that can be classified as purely "technical", intended to convey specific knowledge to those who crossed the seas extensively. These were the actual nautical charts, or *carte da navigare*, which focused solely on the coastline, as the continental landmass was entirely irrelevant for practical maritime navigation. The earliest Venetian representations of the Adriatic belong to this category of medieval nautical charts. Unlike the richly decorated Catalan maps, these Venetian maps were characterized by their rigorous and functional symbolism, an essential and practical design suited to navigational purposes. These charts were often drawn by sailors themselves, based on their seafaring experiences (De La Roncière and Mollat Du Jourdin 1992).

However, a stylistic evolution can be observed within this typology. While the fifteenth-century products featured a simplistic coastal outline, later nautical charts introduced embellishments, such as miniatures of coastal cities. Among these, Venice was disproportionately larger in scale compared to other cities. Allegorical figures, often depicting Venice seated on a throne, and the colouring of Dalmatian islands to emphasize their inclusion in the Venetian domain were also notable additions (Casti Moreschi 1998:48).

A clear example of this evolution can be seen in maps such as Pietro Vesconte's from 1321, Grazioso Benincasa's from 1472, or Gaspare Tentivo's map, which—though much later—adhered to the medieval nautical chart conventions introducing, however, a new west-east orientation which would become the archetype for many Adriatic maps in subsequent centuries (Fig. 4.1). This model persisted until Ludovico Furlanetto's nautical chart from 1784 (see Chap. 1, Fig. 12), whose second edition was printed in 1796, only a year before the fall of the Republic of Venice.

The social and political dimensions were gradually incorporated into later cartographic documents which, while maintaining the traditional structure of nautical charts, added semiotic functions beyond mere

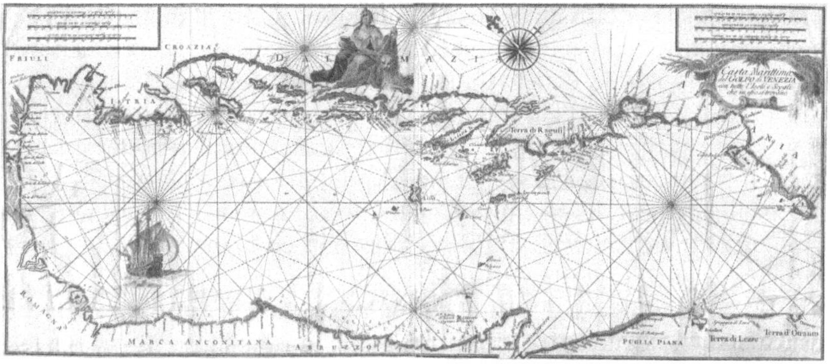

Fig. 4.1 Gaspare Tentivo, *Carta Marittima del GOLFO DI VENEZIA con tutte l'Isole e Scogli che in esso si trovano*, Venice, printed in c. 1740 (Source: Private collection Marco Asta, Bologna)

navigation. They captured and represented the construction of a new geographical space, delineating its limits and forms, which highlighted a distinct and self-contained basin, conceived and governed exclusively by Venetian society. It seems logical that Venice, as the Adriatic's main port and the leading centre for maritime cartographic production of the time, would also be the birthplace of a geometrically conceptualized chorographic archetype. This archetype applied a tailored scale to encapsulate the geopolitical space of the Serenissima within a single sheet.

These documents extended beyond simple coastal descriptions by integrating both physical and human features, such as rivers, mountains, lakes, cities, ports, and fortresses. These elements punctuated the mainland and served as the first instances of "terrestrial" chorographic cartography. Emerging during this period, such innovation bridged the gap between nautical and land cartography by combining maritime and terrestrial elements. The resulting maps fulfilled additional roles, including decorative purposes, geographic information dissemination, and the didactic representation of nautical arts (Bifolco 2020: 20).

Miniatures, coats of arms, flags, architectural elevations, and allegorical figures began to dominate cartographic works in the sixteenth century. This shift coincided with a significant change in the economic policies and lifestyle of Venetian merchants. As they transitioned to managing their affairs from behind desks, delegating the risks and burdens of travel and navigation to others, the cartographic output evolved to meet these new demands. This period saw the emergence of exquisite illuminated collections, carefully bound and organized, grouping maps and charts of known territories to provide detailed information for "armatori" (shipowners) who needed to plan their commercial voyages. Dedication cartouches, often found at the beginning of these luxurious nautical atlases, explicitly reflect this new usage. These materials, commissioned by merchants who had assumed the role of entrepreneurs, were designed to facilitate the planning of trade routes from the comfort of their homes. Typically, such documents portrayed the Adriatic within the broader Mediterranean basin. While adhering to nautical conventions, they began to take on the characteristics of true chorographic maps. Hydrographic, orographic, and settlement features were emphasized to highlight the distinctiveness of specific locations. Notable examples include what is thought to be one of the earliest surviving printed nautical charts, *Exigua hec tabela continet [...] La vera descrittione del Mare Adriatico[...]*, produced by Giovanni Andrea Vavassore in 1539 (and immediately reprinted in 1541), as well as the tables in the nautical atlas edited by Battista Agnese in 1553.

These documents also reveal the vast cultural exchanges, alongside commercial ones, that radiated from the Adriatic coasts into the hinterlands. Examples include Paolo Forlani's charts *Golfo di Venezia* from 1568 (Fig. 4.2) and *Carta del Mediterraneo Orientale* from 1571, as well as the later map by Abbate Lonati (Fig. 4.3), which also features the now traditional northeast-up orientation. The latter, blending nautical and terrestrial elements, stands out as an informative work celebrating the Venetian Dominance represented by the Adriatic Sea stretching from left to right with the central inscription *GOLFO DI VENETIA*. These maps portray the Adriatic as an enclosed basin, almost a private domain, accessible through the Strait of Otranto, labelled here as *la Bocca del Golfo di Venezia*. This is a clear representation of a "Venetian-centric" perception of the Adriatic reality.

Beyond this characterization, another significant category of Venetian cartography must not be overlooked: maps included in *Isolari* (island books) and travel itineraries. These were designed to systematically organize information about insular and mainland areas along coastal lines until they were replaced by modern atlases at the turn of the eighteenth century (Lago 1988).

The *Isolari* were a form of "specialized compendiums", combining maps of individual islands and archipelagos with written descriptions that bridged nautical guides and historical-descriptive cartography. Not only

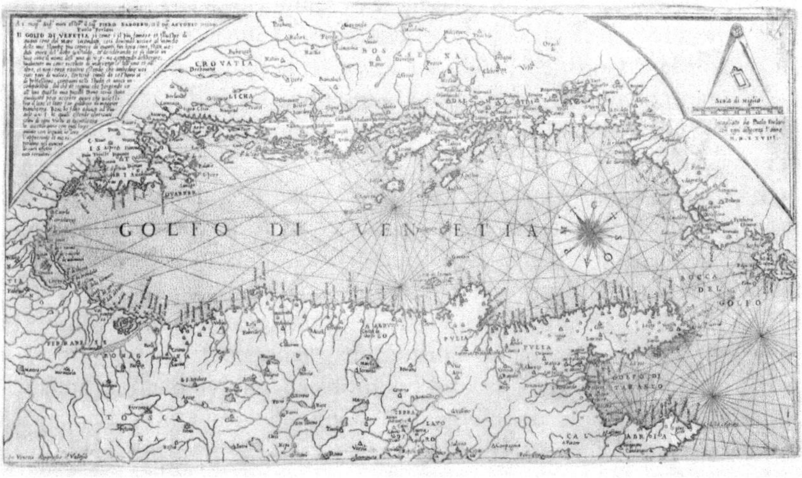

Fig. 4.2 Paolo Forlani, *IL GOLFO DI VENETIA*, Venice, 1568 (Source: Department of Humanities, University of Trieste)

Fig. 4.3 Abbate Lonati, *GOLFO DI VENEZIA*, eighteenth century (Source: Private collection Italico Stener)

did these works provide practical information for merchants and seafarers but also historical notes, classical references, and scholarly inserts. This literary genre, initiated in the fifteenth century by the Florentine Cristoforo Buondelmonti (1420–1422), reached significant prominence in Venice with figures like Bartolomeo Zamberti da li Sonetti (1485), Benedetto

Bordone (1528), Giovanni Francesco Camocio (1571), Tommaso Porcacchi (1572), and Giuseppe Rosaccio (1598), culminating with Vincenzo Maria Coronelli (1696). The *Isolari* became foundational in the development of modern cartography, as no previous tradition had devoted such extensive focus to the insular world (Tonini and Lucchi 2001; Cassi and Dei 1993).

In parallel, travel itineraries offered detailed accounts of routes and places visited. These works not only served as tools for navigation and exploration but also emphasized Venice's role as a bastion of Christianity against the Muslim threat, solidifying its geopolitical position as the gateway to the East. The travel reports and itineraries contained a wealth of information, including maps, illustrations, and detailed descriptions of landscapes, cultures, customs, and traditions encountered along the journey. A prime example of this genre is Giuseppe Rosaccio's *Viaggio da Venezia a Costantinopoli per mare, e per terra, e insieme quello di Terra Santa*, published in Venice in 1598. This work is celebrated for its extraordinary aesthetic value and historical-scientific importance (Conti 2014).

2.2 "*Carte da Terra*"

Of an entirely different nature are the so-called *Carte da terra*, also referred to as *Carte dello Stato da terra e da mar* (maps of the land and maritime state), which were intended for the political-economic management of the Adriatic and its neighbouring territories. This latter designation reveals how the Venetian possessions were inseparably linked to the sea.

The Serenissima's dominion was geographically organized into three entities: the *Dogado*, or the metropolitan territory of the Republic of Venice, encompassing the lagoons between Loreo and Grado, up to the mouth of the Isonzo; the *Stato da terra* or *Dominio di Terraferma* (Mainland Dominion), which included Veneto, Friuli, and parts of Lombardy, with boundaries extending to the Po, the Adda, the Alps, and the Karst Plateau; and the *Stato da mar* or *Dominio marittimo* (Maritime Dominion), a diverse array of possessions stretching from the Adriatic to the eastern Mediterranean, which included Istria, Dalmatia, Corfu, Morea, the Aegean Islands, and Crete (Cessi 1946, 1981). In their various aspects, the control and management of the territories within the Mainland Dominion and the Dogado inevitably led to the creation of countless maps and drawings. Despite their variety, these artefacts assert the political capacity for intervention exercised by the Dominant power.

Even within *Carte da terra*, however, it is possible to distinguish two distinct phases of production, differentiated by the scale of representation, which influenced both their conceptualization and symbolic value. Examples produced at what could be termed a 'regional' scale sought to delineate a politically defined territory and to convey pragmatic information (boundaries, cities, geographical names, etc.), presenting an institutionalized image of the Venetian Dominion. On the contrary, at smaller scales, micro-area maps, characterized by their precision and richness of detail, were of a predominantly instrumental nature. They served as tools for administration and territorial understanding under public authority or were aimed at the management of private property (Sereno 2007).

In its dual public and private dimensions, this terrestrial cartography introduced a revolution in Venetian cartographic metrics: a shift from recording linear space—focused exclusively on the coastline—to areal metrics—representing the territory in its entirety. This transition marked the emergence of a new spatial perception tied to the evolving political role of the territory within the Dominions and the necessity of understanding the regions of the Mainland under the Venetian administration. The transformation in representation reflected a broader geopolitical paradigm shift, where the maritime dimension (the sea) was increasingly integrated with the terrestrial component (the land), transitioning from maritime dominance to territorial control.

Several surviving examples attest to this pivotal transition. Among them is Stefano Scolari's map, which prominently highlights the boundaries of Venetian possessions and the geopolitical configuration of the northern Adriatic at the border with Austrian (Habsburg) territory (Fig. 4.4). This map is an exceptionally rare and aesthetically striking artefact. It delineates the shores of the upper Adriatic, from the mouth of the Po to the Dalmatian islands facing Zadar, and encompasses a territorial scope stretching from the Carnic tributaries of the Tagliamento to the Canal Valley, Ljubljana, and the mountains north of Zagreb. Even upon initial examination, the "political" nature of this document is immediately evident. It aims to depict the Serenissima's dominions and celebrate its prominence. A red watercolour line clearly demarcates Venetian territories, while the inscriptions "Italy ends here" emphasize a cultural boundary between Venetian lands and Austrian domains. Additionally, within the sea, the map features a sumptuous allegorical representation of Ducal Venice, placed alongside

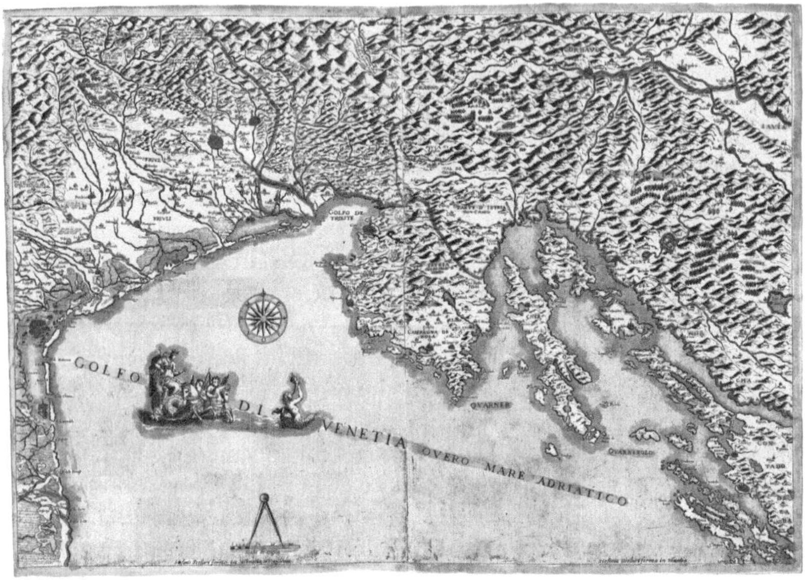

Fig. 4.4 Stefano Scolari, *GOLFO DI VENETIA OVERO MARE ADRIATICO*, Venice, the end of the sixteenth century (Source: Private collection Italico Stener)

the grand inscription *GOLFO DI VENETIA OVERO MARE ADRIATICO*. This inscription almost serves as an official declaration, emphasizing the natural continuity of jurisdiction between land and sea (Salgaro 2007).

Other notable examples in this regard include engravings stemming from the Gastaldian tradition, printed in the Venetian workshops of Matteo Pagano, such as the large map depicting parts of Italy and the Dinarides region from 1546, now housed in the Vatican Apostolic Library. Another is the map of the northern Adriatic printed from Paolo Forlani's plates at the "Sign of the Column" in 1566, which includes the northern Adriatic regions and the coastline stretching from Ancona to the island of Korčula. Although neither of these maps place Venice at the centre of the composition, they nonetheless convey the lagoon city's regional hegemony and its dominions.

In addition to these general-purpose documents, which offer a comprehensive overview, it is important to note that starting in the first half of the fifteenth century, Venice systematically seized control of mainland cities up to the Adda River through a series of wars. As a result, the Republic came into possession of the cartographic archives of the local lordships. These detailed maps of the various provinces brought under Venetian control (such as Padua, Treviso, Vicenza, Verona, and Feltre) were incorporated into the Republic's archives during territorial annexations. Throughout the fifteenth century, these maps were supplemented by those created by Venetian military engineers. Numerous derivative maps from this period attest to the continuous effort to update the Republic's geographic knowledge through field data collection. However, the Republic did not consider this rich cartographic heritage sufficient for effective governance. On February 27, 1460, the *Consiglio dei Dieci* (Council of Ten) issued a decree mandating the compilation of a comprehensive cartographic corpus of the mainland, with maps for individual provinces to be stored in the *Cancelleria* (Council of Ten Chamber). The decree stipulated that this collection would ensure "that anyone who wishes to be informed and take measures regarding the territories of the *Dominio* may have accurate and detailed knowledge without consulting anyone else"[1] (Cavazzana Romanelli and Casti Moreschi 1983: 8).

Alongside this extensive body of official information, produced by the Republic's complex administrative apparatus—comprising various state bodies such as the *Provveditori*, *Sovrintendenti*, *Savi*, *Periti*, and *Magistrati*—there was also a wealth of cartographic material commissioned by private individuals to meet diverse needs and interests. This included maps requested by travelling merchants for their trade activities, by farmers delineating land for cultivation, by Venetian patricians seeking to celebrate the status of their estates, and by landowners planning transformations such as cultivation, infrastructure development, or construction. It was not uncommon for this type of "minor" and non-institutional cartography to move beyond its private scope in cases of disputes, becoming legal evidence and testimony to reality. This transition enriched the Serenissima's documentary heritage even further, as private cartography acquired a legal and evidentiary value (Casti Moreschi 1999; Vantini 2007).

[1] "Pro omni bono respectu providendum est habere in Cancellaria nostra aut Camera Consilii nostri decem in vera pictura formam et exemplum omnium civitatum terrarum castellorum provinciarum et locorum nostrorum, ut quicumque volens consulere et providere super predictis habeat veram et particularem noticiam ad concilium et non ad opinionem alicuius" (ASVe, *Consiglio dei Dieci*, *Misti*, Registro 15, 1454–1459, carta 197).

That corpus of representations, produced between the sixteenth and eighteenth centuries, was characterized by an uncodified cartographic tradition. It depicted areas of varying size, relied on a certain empiricism in surveying measurements on the ground, and belonged to a well-established Venetian tradition. At the same time, it remained a work in progress, evolving in response to the shifting styles of the figurative arts of the period and the individual preferences of cartographers, which changed continually over time. Before the eighteenth century, it was common practice to document territorial information through detailed written reports, accompanied by cartographic images that provided precise descriptions of places, features, and challenges. The absence of a rigid representational model allowed considerable latitude for individual cartographers' interpretations, resulting in documents of varying formats, scales, and conceptions; views and perspective drawings alternated with micro-area maps, sketches, and fully fledged regional maps. This diversity led to a rich cartographic production, both quantitatively and qualitatively, serving the political-administrative needs of the Venetian state as well as the interests of private individuals.

The archives[2] of Venice contain abundant records that attest to the extensive activity of civil engineering, including the management of water systems, forests, and landed estates, urban planning, border control, and countless other aspects of geo-physical and anthropic interventions on the territory.

Among the many examples, we can highlight the management of water resources, which was a constant concern for the Venetian government. Over the centuries, water governance became increasingly stringent, driven by the critical need to protect the city and the lagoon from the effects of both the sea and the rivers. The rivers, which flowed abundantly into the Gulf, posed a significant risk of silting the coastal zones, thereby endangering the transportation routes within the mainland territories (Bevilacqua 1970).

This concern was so paramount that, around the fifteenth century, a dedicated office, the *Savi alle Acque*, was established. Together with the *Esecutori alle Acque* and the *Collegio Solenne alle Acque*, these entities formed the renowned *Magistrato alle Acque*, the body responsible for regulating rivers, shores, and the lagoon. It was within this framework that

[2] Reference is made to the current funds present in the State Archives (ASVe), in the Correr Museum, or in the Marciana Library.

various cartographic documents were produced, differing in composition, scale, and detail. These maps were created by engineers and surveyors commissioned by the Venetian state to report on the status of individual hydrographic basins and address related issues, encompassing technical, functional, physical, political, human, and structural aspects (Caniato 1993). A notable example of such work is the series of maps by Cristoforo Sabbadino, which document the extensive operations undertaken by the Serenissima, starting in the fourteenth century and significantly intensifying from the late fifteenth century. These operations aimed to divert a substantial portion of the rivers (such as the Brenta, Bacchiglione, Piave, and Sile) that flowed into the lagoon to prevent its silting. Another significant piece is the large manuscript map by Cristoforo Sorte, now preserved at the State Archives – War Archives (Staatsarchiv-Kriegsarchiv) in Vienna (see Fig. 4.5). As the first official surveyor for uncultivated lands, Sorte, drawing on the surveys conducted by hydrographers and topographers like Sabbadino, produced a detailed summary on the hydrographic situation of the Venetian Dominions in the late sixteenth century. This map accurately depicts the major rivers, tributaries, streams, canals, lakes, ports, bridges, beaches, and other elements of the intricate water system (Salgaro 2012).

Venetian interest extended far beyond water-related matters, encompassing various aspects of territorial management, with particular attention devoted to the control and administration of forests (Casti Moreschi 1994). This domain generated a significant body of cartographic material, notable for its representational clarity and visual richness. Forests were regarded as a resource to be preserved, protected, and defended, with severe penalties, including the death penalty, imposed for serious violations. The strict regulation of forest resources by the Venetian Senate and the Council of Ten was primarily motivated by the unique nature of the lagoon city, much of which was built on wooden piles, as well as by the need to maintain the structural integrity of its urban fabric, poised between land and sea. Moreover, there was an acute awareness of the essential role of timber in sustaining the Venetian fleet, a key instrument of defence and expansion, as well as in supporting urban construction, fulfilling energy needs, and meeting everyday domestic and industrial demands (Casti Moreschi and Zolli 1988).

The issue of forest management was of concern to all—both private citizens and institutions—and thus received significant attention from the authorities due to its far-reaching implications. Reckless deforestation or

Fig. 4.5 Cristoforo Sorte, *Dissegno dilla Patria dil Friuli...*, Venice, 1590 (Source: Staatsarchiv-Kriegsarchiv, Vienna, KA/KS, B VII a 167)

misappropriation of this precious natural resource could destabilize the entire territorial and environmental balance. Similarly, its utilization was inextricably linked to the need for an efficient, well-maintained hydrographic network. The transport of timber, whether from mountainous or lowland forests, relied heavily on waterways. Logs were floated to Venice along rivers or transported on rafts, needing meticulous planning. Before any felling occurred, a reconnaissance mission was carried out by experts tasked with evaluating the economic value of the timber and determining the methods for cutting the trees. Certain forests were designated as public property and thus protected by laws and regulations, with officials overseeing the operations during timber harvests. One notable example is the detailed exploitation plan for the forests above the Lumiei stream near the village of Sauris, prepared in 1752 by the surveyor Giuseppe Fabris. Several cartographic documents from this project have survived, providing graphically precise illustrations of even the smallest details (Bianco 2001: 18) (see Fig. 4.6).

Each of these documents lends itself to interpretation on multiple levels, as they contain diverse types of information that serve purposes beyond a

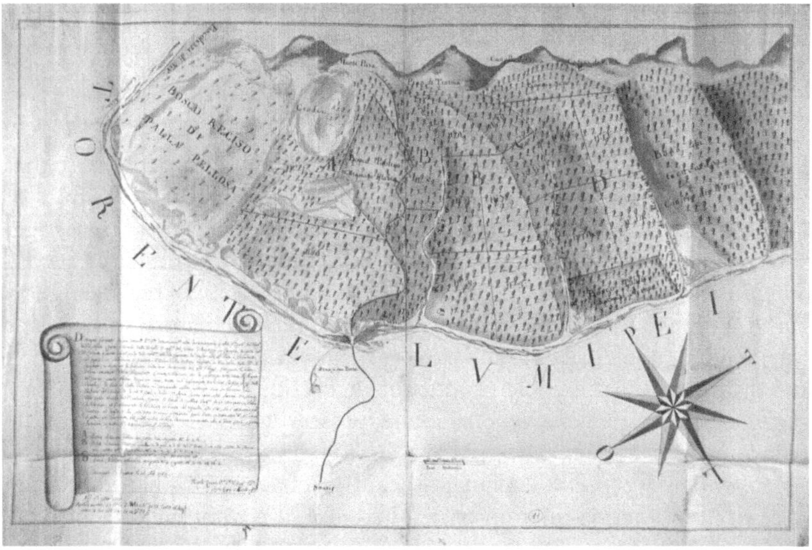

Fig. 4.6 Francesco Leschiutta, *TORENTE LVMIEI*, 1769 (Source: ASVe, Provveditori Sopra Boschi, b. 170, dis. 11)

single, exclusive function. They represent a reality of coexistence, where the data described and mapped offer a rich and complex picture of the interplay between natural forces and human intervention. These records often evoke the history and events of places far removed from the Venetian Lagoon.

For instance, the manuscript map titled *Disegno del Posto della Chiusa* created by Giacomo Spinelli in 1703 serves as a significant historical document and is a perfect example of the interpretation as mentioned above. This author was a surveyor listed in the "Catalogo delli periti, calculatori et agrimensori pub[blici]", a professional register established by the government of Udine in 1661. This register included the most distinguished experts in land surveying and cartography of the region (Lago and Rossit 1988: 93). The document (see Fig. 4.7) not only demonstrates Spinelli's deep and accurate understanding of the areas represented, but also highlights the technical skill involved in its creation and the considerable attention that the Serenissima devoted to its mountainous territories.

This map highlights the need for certain restorative works along the Fella River, including the construction of artificial barriers and the establishment of lookouts and garrison posts along the mountain's slope to enhance the security of the "Castello della Chiusa di Venzon", located six miles from Venetian Ponteba on the border of the Austrian State in Friuli (Bianco 1985; Cacciavillani 1991).

3 Conclusion

The analysis presented here seeks to highlight the significant influence that Venetian cartography exerted on the political and territorial dynamics of the Serenissima, as well as its impact across various domains of knowledge. A blend of science, imagery, and metaphor, Venetian cartography served multiple purposes from its inception: as a reduced graphical representation of the land, as a tool for political propaganda, as a celebratory and decorative image in noble palaces, and as an instrument for navigation. This rich and distinctive cartographic production, reflecting diverse needs, accompanied the history of the Venetian State, preserving its documentary value even after its fall.

It is widely accepted that a map or nautical chart is not a faithful mirror of reality but a representation of it. Therefore, it cannot be judged solely based on its degree of accuracy and adherence to reality but should be viewed as an intellectual appropriation and construction of territory by a community. In this sense, maps and nautical charts acquire a complex and

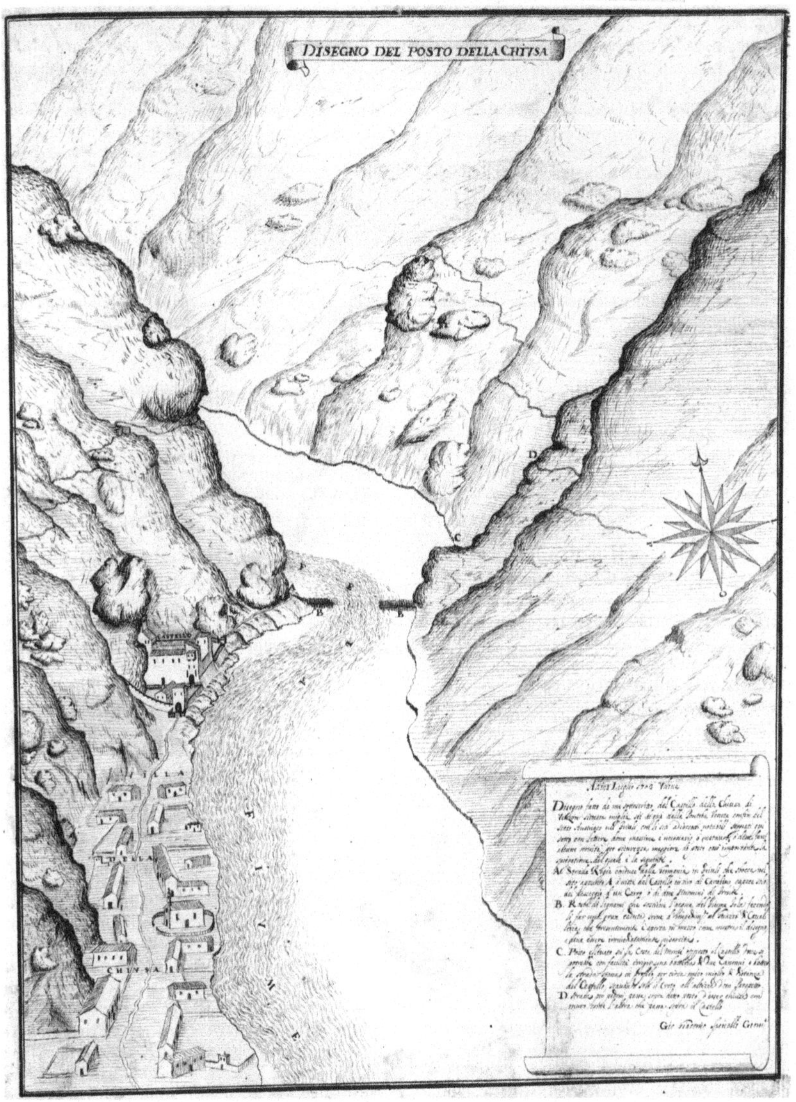

Fig. 4.7 Giacomo Spinelli, *DISEGNO DEL POSTO DELLA CHIUSA*, 1703 (Source: ASVe, Raccolta Terkuz, dis. 45 A)

historically contextualized "polysemic" character, where precision and objective correspondence to reality are just some of the variables to analyse. Moving beyond the paradigm of precision and aesthetic criteria, every aspect of this polysemy becomes essential for decoding the forms, functions, utilities, roles, and strategies implied and embodied by each cartographic representation.

Thus, every map and nautical chart emerges as a complex cultural artefact where each element—lines, cartouches, symbols, names, and colours—intertwines to make the perceived space tangible, visible, and material. The extensive Venetian cartographic heritage, which can be categorized into the two main streams of "technical cartography" and "symbolic cartography", constitutes a significant documentary sequence in terms of quantity, quality, and continuity. It provides a valuable basis for semantic analysis of the documents. Beyond being an expression of political, economic, and military power, cartographic documents, when analysed in terms of their intrinsic and extrinsic characteristics, allow us to trace the technical and scientific evolution of cartography and its dissemination as a tool for territorial knowledge and management. The need to understand space represents not merely a formal act—of which maps and nautical charts offer an excellent visual summary—but the beginning of a process in which maps and charts serve to dominate nature, regulate rights, certify ownership, plan interventions, and manage conflicts.

Respecting the various methodologies and specific interpretative approaches of individual scientific disciplines, each of these old documents offers valuable insights into the processes of constructing territorial identities and geographic imaginaries. In the case of Venice, this process of identification is particularly evident, first in the vision of its sea (the Adriatic) and later in its territories (the mainland).

References

Almagià R (1929) Monumenta Italiae cartographica: riproduzioni di carte generali e regionali d'Italia dal secolo XIV al XVII. Istituto Geografico Militare

Astengo C (2000) La cartografia nautica mediterranea dei secoli XVI e XVII. Erga

Barsanti D (1985) Senso e significato dei cimeli geocartografici. Geografia 8:55–58

Bevilacqua E (1970) La cartografia storica della Laguna di Venezia, "Mostra storica della laguna veneta". Stamperia di Venezia

Biadene S (1990) Carte da navigar: portolani e carte nautiche del Museo Correr 1318-1732. Marsilio Editori

Bianco F (1985) Comunità di Carnia. Le comunità di villaggio della Carnia (secoli XVII-XIX). Casamassima

Bianco F (2001) Nel bosco. Comunità alpine e risorse forestali nel Friuli in Età moderna. (secoli XV-XX). Forum

Bifolco S (2020) Mare nostrum. Cartografia nautica a stampa del Mar Mediterraneo. Edizioni Antiquarius

Boria E (2020) Storia della cartografia in Italia dall'Unità a oggi: tra scienza, società e progetti di potere. UTET

Braudel F (1976) Civiltà e imperi del Mediterraneo nell'età di Filippo II. Einaudi

Cacciavillani I (1991) La confinazione veneziana con gli imperiali. Signum Verde

Caniato G (1993) La via del fiume. Dalle Dolomiti a Venezia. Cierre Edizioni

Caraci G (1959) Di alcune carte nautiche anonime che si vorrebbe attribuire a Girolamo da Verrazzano, Istituto Geografico Militare

Cassi L, Dei A (1993) Le esplorazioni vicine: geografia e letteratura negli "Isolari". Rivista Geografica Italiana, Annata del Centenario 100:205–269

Casti Moreschi E (1993) Cartografia e politica territoriale nella Repubblica di Venezia (secoli XIV-XVIII). In: Institut Cartografic Catalunya & Universidad Autonoma Barcelona (eds) La cartografia italiana. Institut Cartografic de Catalunya, pp 81–101

Casti Moreschi E (1994) Cartografia e politica territoriale: i boschi a Venezia. Storia Urbana 69:105–132

Casti Moreschi E (1997) Rappresentazione e pratica denominativa: esempi dalla cartografia veneta cinquecentesca. In: Galliano G (ed) Rappresentazioni e pratiche dello spazio in una prospettiva storico-geografica, Centro Italiano per gli Studi Storico-Geografici, pp 109–138

Casti Moreschi E (1998) L'ordine del mondo e la sua rappresentazione. Unicopli

Casti Moreschi E (1999) L'Adriatico rappresentato. In: Turri E (ed) Adriatico Mare d'Europa. La geografia e la storia. Rolo Banca 1473, pp 46–51

Casti Moreschi E, Zolli E (1988) Boschi della Serenissima. Storia di un rapporto uomo-ambiente. Arsenale Editrice

Cavazzana Romanelli F, Casti Moreschi E (1983) Laguna, lidi, fiumi: esempi di cartografia storica commentata. Ministero per i Beni Culturali e Ambientali, Archivio di Stato di Venezia-Regione Veneto, Dipartimento per l'Informazione-IRRSAE Veneto. Tipografia Helvetia

Cessi R (1946) La Repubblica di Venezia e il problema adriatico. Edizioni scientifiche italiane

Cessi R (1981) Storia della Repubblica di Venezia. Giunti

Conti S (2014) Il viaggio verso l'oriente meraviglioso e l'opera di Giuseppe Rosaccio. In: Siniscalchi S (ed) Studi in onore di Vincenzo Aversano. University of Salerno, Gutenberg Edizioni, pp 453–486

Cucagna A (1964) I Friuli e la Venezia Giulia nelle principali carte geografiche dei secoli XVI, XVII e XVIII. Catalogo ragionato della Mostra storica di cartografia. "Atti del XVIII Congresso Geografico Italiano", vol III. Tipografia Moderna

Dai Prà E (2010) Il paesaggio liminare nella cartografia storica tra metafora e progetto. Bollettino A.I.C., 139-140/2010, 167-179. EUT Edizioni Università di Trieste
De La Roncière M, Mollat Du Jourdin M (1992) I Portolani: Carte nautiche dal XIII al XVII secolo. Bramante Editrice d'Arte
Dodge M, Kitchin R, Perkins CR (2009) Rethinking maps. Routledge
Edney MH (2019) Cartography: the ideal and its history. University of Chicago Press
Farinelli F (1992) I segni del mondo. Immagine cartografica e discorso geografico in età moderna. La Nuova Italia
Farinelli F (2003) Geografia: un'introduzione ai modelli del mondo. Einaudi
Farinelli F (2009) La crisi della ragione cartografica Geografia: un'introduzione ai modelli del mondo. Einaudi
Ferro G (1992) Carte nautiche dal Medioevo all'età moderna. Edizioni Colombo
Harley JB (1989) Deconstructing the map. Cartographica 26(2):1–20. https://doi.org/10.3138/E635-7827-1757-9T53
Heffernan MJ (2007) The European geographical imagination: Hettner-Lecture. Steiner
Ivetić E (2007) Per una periodizzazione della storia dell'Adriatico orientale. Atti del Centro di Ricerche storiche di Rovigno 37:161–180
Lago L (1988) Imago Adriae. La patria del Friuli, l'Istria e la Dalmazia nella cartografia antica. La Mongolfiera
Lago L (1989) Theatrum Adriae: dalle Alpi all'Adriatico nella cartografia del passato (secoli X-XVIII): mostra di cartografia antica, Museo provinciale di storia ed arte, Gorizia Borgo Castello 23 giugno-9 ottobre 1989. Lint
Lago L, Rossit C (1988) Theatrum Fori Ivlii. La Patria del Friuli ed i territori finitimi nella cartografia antica sino a tutto il secolo XVIII, Volume Secondo. Il Seicento e il Settecento. Lint
Licini P (2000) Formula picturarum "ciberbetica". Geostorie. Bollettino e Notiziario del Centro Italiano per gli Studi Storico-Geografici 8(1–2):25–55
Luzzatto G (1995) Storia economica di Venezia dall'XI al XVI secolo. Marsilio Editori
Marinelli G (1881) Saggio di Cartografia della regione veneta, Monumenti Storici pubblicati dalla Regia Deputazione Veneta di Storia Patria, serie IV, vol I, Miscellanea I. Tipografia Naratovich
Milanesi M (2008) Cartografia per un principe senza corte. Venezia nel Quattrocento. Micrologus 16:189–216
Milanesi M (2012) Nelle stanze di palazzo. Venezia, Firenze, qualche confronto. In: Salgaro S (ed) Cristoforo Sorte e il suo tempo. Pàtron editore, pp 97–117
Pavanetto L (2019) Mappe ed esploratori sulle rotte della Serenissima. Dario De Bastiani Ed
Quaini M (1994) La carta geografica. Un racconto speculare. Rivista Geografica Italiana 101:319–326

Salgaro S (2007) Cartografi e cartografia come strumenti di controllo e gestione territoriale nella Repubblica di Venezia. In: Valerio V (ed) Cartografi Veneti. Mappe, uomini e istituzioni per l'immagine e il governo del territorio. Editoriale Programma. pp 33–45

Salgaro S (2012) (ed) Cristoforo Sorte e il suo tempo. Pàtron editore

Schulz J (1990) La cartografia tra scienza e arte: carte e cartografi nel Rinascimento italiano. Panini Editore

Selva O (2016) Lo stato della cartografia veneziana tra XVI e XVIII secolo: emblema di potere e strumento di pianificazione territoriale. Bollettino dell'Associazione Italiana di Cartografia 148:69–87

Sereno P (2007) Ordinare lo spazio, governare il territorio: confine e frontiera come categorie geografiche. In: Pastore A (ed) Confini e frontiere nell'età moderna un confronto tra discipline. Franco Angeli, pp 45–64

Sturani M L (2008) Le rappresentazioni cartografiche nella costruzione di identità territoriali: materiali e spunti di riflessione dalla prospettiva della storia della cartografia. In Organizzazione del potere e territorio. Contributi per una lettura storica della spazialità. Franco Angeli. pp. 189–213

Tonini C, Lucchi P (2001) Navigare e descrivere. Isolari e portolani del Museo Correr di Venezia XV-XVIII secolo. Marsiglio Editore

Valerio V (2007) Cartografi Veneti. Mappe, uomini e istituzioni per l'immagine e il governo del territorio. Editoriale Programma

Vantini S (2007) Periti, agrimensori, notai: cartografia e cartografi minori tra amministrazione periferica e Magistrature centrali negli ultimi due secoli della Repubblica veneta. In: Valerio V (ed) Cartografi Veneti. Mappe, uomini e istituzioni per l'immagine e il governo del territorio. Editoriale Programma. pp 19–32

Wood D, Fels J (1992) The power of maps. Guilford Press

CHAPTER 5

Symbolism of Artistic Elements on Nautical Charts of the Adriatic Sea

Lena Mirošević and Dubravka Mlinarić

1 Introduction

The representation of cognitive geoinformation is as old as human history itself. The design of such representations—especially the visual presentation of geographical spatial data—is considered a cultural universal (Blaut et al. 2003). Schematic or cartographic representations were not only reflections of knowledge about new spaces but also instruments of power. It is not only a question of what is shown (or omitted) on the map but also who created it and for what purpose. The production of maps and the dissemination of cartographic knowledge developed parallel with advances in techniques and technologies for visualizing geographical data, especially by the era's new geographical discoveries. Nautical cartography and nautical charts played a crucial role as the primary source of information for

L. Mirošević (✉)
Department of Geography, University of Zadar, Zadar, Croatia
e-mail: lmirosev@unizd.hr

D. Mlinarić
Institute for Migration Research, Zagreb, Croatia
e-mail: dubravka.mlinaric@imin.hr

J. Faričić, T. Marelić (eds.), *Early Modern Nautical Charts of the Adriatic Sea*, https://doi.org/10.1007/978-981-96-8046-7_5

constructing a global and geographically comprehensive understanding of the world. Such a perspective had been available for the Mediterranean region since the late Middle Ages through portolan charts. These charts, designed to aid navigation, were based on data gathered during maritime voyages and documented coastal features, islands, and the contours of landmasses along the oceans. Consequently, the production of cartographic knowledge was organized, curated, controlled, and redistributed according to specific procedures to a considerable extent until the early modern period (Harley 1988a). In this context, maps served as communication tools, connecting various actors through both textual and extratextual graphic forms. They played a crucial role as a medium that served state interests by aiding in the production and representation of spatial knowledge, while also reinforcing territorial control and projecting power over space (Panofsky 1970; Brunn and Dodge 2017; Wintle 1999; Mlinarić and Miletić Drder 2017; Duzer 2021). The art of cartography in the early modern period followed specific norms. This was primarily because the visual representation of geographical space needed to avoid abstraction or deviation from established practices associated with cartographic signs. By doing so, it could effectively maintain its broader communicative function (Mlinarić and Miletić Drder 2017; Faričić et al. 2023a). On European maps of the sixteenth and seventeenth centuries, this can be seen in the practice of geostrategic and economic (business) secrecy or a combination of both. In this context, the map serves as a medium of communication, conveying data about and in geographical spaces, and, most importantly, providing a conceptual basis for planning and taking action (Cosgrove 2008; Harley 1988b). To fully understand the content and meaning of a map, it is essential to consider the social, historical, and technical context and processes from which it originated and on which it exerts influence. This requires an interdisciplinary approach to the study of the map—not just as an isolated object, but as a product of specific technical and social processes and as a catalyst for broader social dynamics as it circulates within society. Apart from serving their practical purpose of depicting spatial relationships and information, a significant emphasis is placed on the aesthetics of cartographic design. This effort aims to create harmony and balance in map representation. Additionally, maps often feature decorative elements that enhance both their artistic appeal and functional utility (Rees 1980; Robinson 1982; Faričić et al. 2023a). The harmony between the graphic design of the map as an artistic representation of space and its geocartographic informativeness lies in the balance between scientific accuracy and artistic creativity within the cartographic content (Ribeiro and Caquard 2018; Cartwright 2009; Krleža and Mlinarić 2022). At the same time, graphic elements inspired by Christian iconography frequently appeared on early modern maps of the Mediterranean. As the cultural, social, and economic crossroads of south-eastern Europe, the Adriatic was characterized by a variety of graphic elements. These elements reflect the region's role as a meeting point for different ideas, practices, and interests.

2 Research Framework, Sources, and Methodology

This study examines the role of cartographic decorative and imagological elements in early modern representations of the Adriatic Sea. Its aim is to evaluate the communicative capacity of these cartographic representations as carriers of ideological, political, cultural and other symbolic messages. Furthermore, the focus is on the multiple interpretations and meanings derived from the symbolic power of extratextual communication (Panofsky 1970; Mlinarić and Miletić Drder 2017). A qualitative comparative analysis of the symbolic elements in cartographic representations of the Adriatic Sea from various, mainly Venetian, sources was carried out. This analytical comparison aims to uncover the range of messages these nautical charts and maps convey. The study examined selected early modern nautical charts and other types of maps (but always related to naval, commercial, political, or any other aspect of seafaring) depicting the Adriatic from different sources, including Venetian and French (Table 5.1). By employing a comparative analysis from both synchronic and diachronic perspectives and utilizing techniques of decomposition and deconstruction, we can interpret these nautical charts and maps in terms of their symbolism. This includes an examination of the code and the role of iconographic (decorative) and imagological cartographic elements (symbols) such as compass roses, vedute, images of patron saints, and depictions of animals, monsters, and humans (Mitchell 1986; Harley 1992). These two criteria—decorative and utilitarian purposes—are often interwoven through the selection and use of sophisticated symbolic signs (cartographic signs with a specific explicit or hidden meaning). In deconstructing cartographic representations, the following analytical key pillars are identified: the rhetoric of decorative elements in their different forms, the presence of hidden or

Table 5.1 Maps and nautical charts depicting the Adriatic Sea used in the research

Author	*Title – content*	*Place and year of production*	*Institution where it is kept and call number*
Unknown author	Nautical chart of the Adriatic Sea (in Dauphin Atlas)	Dieppe, c. 1538–1546	Koninklijke Bibliotheek, Nationale bibliotheek, The Hague
Unknown author	Nautical chart of the Adriatic Sea (in Vallard Atlas)	Dieppe, 1547	Berkeley Digital Library, Henry E. Huntington Library and Art Gallery, San Marino (California, SAD), sign HM 29, f. 14
Joan Martines	Nautical chart of Aegean Sea and eastern Mediterranean	Messina, c. 1578	The Huntington Library, Art Museum, and Botanical Gardens, Library Collections, Maps and Atlases, Portolans, San Marino, CA, USA, mssHM 33
Willem Barentsz	*Tabula Hydrographica, In qua Italiae, orae maritimae; Item Venetiae, Istriae, Dalmatiae, Slavoniae, Graeciae…*	Amsterdam, 1595	The Barry Lawrence Ruderman Map Collection, Stanford University Libraries, Tooley, A-D:85, 268; K-P:10
Niccolò Guidalotti	Nautical chart of the Adriatic Sea	Venice, 1646	Bibliteca Nazionale Marciana, Venice, It. IV, 10=5062
Vincenzo Maria Coronelli	*Golfo di Venezia*	Venice, 1688	National and University Library, Map and Atlas Collection, Zagreb, S-JZ-XVII-56.
Vincenzo Maria Coronelli	*Ristretto della Dalmazia*	Venice, 1690	Maritime and Historical Museum of the Croatian Littoral, Rijeka Call number PPMHP 122117
Gasparo Tentivo and Francesco Felice Polanzani	*Carta Marittima del Golfo di Venezia*	Venice, c. 1740	Istituto Geografico Militare, Florence
Joseph Roux	*Carte de la mer Mediterranee*	Marseille, 1764	National and University Library, Map and Atlas Collection, Zagreb S-JZ-XVIII-116

misrepresented content, and the hierarchy of the represented elements (Harley 1992; Faričić 2007; Cosgrove 2008; Fuerst-Bjeliš 2012; Mlinarić 2012; Šakaja 2015; Monmonier 2018). All of these concepts were interwoven and unified in the cartographic representations, probably to help the mapmakers convey their message more effectivsely and subtly about the space they were interested in.

3 The Rhetoric of Decorative Elements

As media, maps served as significant but subtle means of communication between different actors and promoted state interests, whether through the production and presentation of knowledge about a particular area or the assertion of territorial control and power. Nautical charts, aside from their aesthetic appeal, can communicate a wide variety of messages through their graphic elements. They can provide information about the geographical features of maritime or land areas surrounding the coastline, reflect the cultural and political affiliations of a specific region, and convey the perspectives of the power or interests that influenced the cartographer. These factors can also support, shape, or alter the creation of the map. The cartographic repertoire of decorative elements on the maps thus resulted from the expectations of the general public on the one hand and the clients, patrons, and sponsors on the other. One of the particularly conspicuous influences on the Venetian nautical charts of the Adriatic was sometimes manifested in the persuasive policy of the Serrenissima as one of the most powerful political forces on the Adriatic.

3.1 Geostrategic Rhetoric of the Map: Historical-Geostrategic Context of the Production of Nautical Charts of the Adriatic in the Light of Venetian Expansionist Policy

As part of its overseas possessions, the centuries-long Venetian rule over much of the northeastern Adriatic coast profoundly shaped the region's political, economic, and cultural landscape. Venice aimed to establish itself as the dominant power in the Adriatic by asserting a monopoly over shipping and trade, often positioning itself as a maritime force with the authority to control the entire sea (Faričić et al. 2023b; Ivetic 2019). The political ideology of the Venetian Republic is reflected in the iconography and symbols found on early modern nautical charts of the Adriatic. These cartographic elements highlight Venice's political ambitions in the region. The messages conveyed through these charts were often politically motivated and shaped by the circumstances of the sixteenth, seventeenth, and early eighteenth centuries. The Adriatic was indeed a theatre of contacts at the time, many of which were conflictual, especially between three great powers: the Venetians, the Habsburgs, and the Ottomans. The situation in the western part of the Adriatic was further complicated by the diverse military and economic interests of the politically fragmented Appenine peninsula (Ivetic 2019). During this time, Venice engaged in three wars against the Ottoman Empire: the Cretan War (1645–1669), the Morean War (1684–1699), and the Second Morean War (1714–1718). Although Venice ultimately lost its territories in Crete and the Peloponnese, the Republic successfully expanded its holdings in the hinterland of the north-eastern Adriatic coast. In response to these conflicts, Venice made efforts to map both its existing and newly acquired overseas territories in the Adriatic with increasing accuracy (Mayhew 2008). In addition to legal and military measures, Venice sought to preserve its supremacy in the Adriatic across all domains, aiming to solidify its image as the mistress of the sea. On their maps, Venetian cartographers referred to the Adriatic as the "Gulf of Venice" (*Golfo di Venezia*), a term that was later adopted by other European cartographers who used Venetian maps as models for their own creations (Faričić et al. 2023b). This intellectual discourse is particularly evident in the work of Vincenzo Maria Coronelli, the leading Venetian cartographer of the late seventeenth century. Coronelli produced numerous maps of the Adriatic coastline, which became a key reference for many European cartographers. As the official cartographer to the Venetian Senate, his work can be interpreted as a deliberate effort by the Venetian state to project its political power. Coronelli's cartography exemplifies Venetian imperial discourse, transforming geography into a tool of state propaganda (Stouraiti 2002). His maps serve as vital documents that not only capture the major historical milestones in the Republic's political narrative—ranging from territorial conquests to the depiction of the Ottoman Empire—but also embody the Republic's ambitious vision for the future. On the nautical chart of the Adriatic Sea (1688), for instance, the Adriatic Sea is named the "Gulf of Venice", with the additional note "formerly known as the Adriatic"(Golfo di Venezia olim Adriaticum Mare) (Faričić et al. 2023b) (Fig. 5.1). This choice of naming illustrates a desire to reshape historical perceptions and assert a distinct identity. The laurel cartouche prominently highlights Venice as the key feature, representing both the capital and ruler of the Venetian Gulf (*Capital, e dominante del*

Fig. 5.1 Coronelli's map of the Adriatic Sea, Venice, 1688 (Source: National and University Library, Map and Atlas Collection, Zagreb, Call number S-JZ-XVII-56)

Golfo!). It is surrounded by vedute of smaller (inferior) coastal and island cities along the eastern and western Adriatic coasts, which symbolize Venice's dominance and influence. This concept of superiority is further emphasized by the size of the vedute; Venice stands out prominently, while the other cities around the Adriatic are shown uniformly.

3.2 *Confessional Rhetoric: Highlighting the Use of Religious Imagery to Define Space*

In addition to providing geographical information, early modern nautical charts of the Adriatic also conveyed cultural, political, and religious messages, reflecting the Mediterranean's complex historical and political dynamics. One of the subtlest religious symbols on these nautical charts is the specific graphic element within the compass rose, an almost ubiquitous feature due to its role in indicating orientation. The compass rose evolved from the earlier wind roses, which marked the directions from which the winds blew, often labelled with the initial letters of their names. In addition to these wind-related letters, religious symbols were employed to mark cardinal and inter-cardinal directions. For example, the north was often represented by a simplified arrowhead or a fleur-de-lis, while the sign of the cross indicated east (Faričić et al. 2023a). The fleur-de-lis, in particular, carried religious and heraldic significance: it was associated with the Virgin Mary and the Holy Trinity in a religious context and with the royal houses of France and England in a heraldic one. The use of the sign of the cross for the cardinal point east is closely tied to Christianity, as it points towards Jerusalem (see Chap. 6, Fig. 6.2). Both the lily flower and the cross symbolize a strong cultural and religious identity, reflecting the intertwined nature of different communities in the Mediterranean region (Faričić et al. 2023a).

In the context of the Venetian spatial engraving, the cartographic representation of Coronelli's Venetian Dalmatia (see Chap. 6, Fig. 6.6b) is particularly interesting. It is dominated by the allegorical depiction of the triumphal chariot of Venetian dominance under the Christian protection, suggesting a call for political and confessional unity between Venice and Venetian Dalmatia. The culmination of establishing hegemony in the Adriatic and Ionian areas was reached with the cartographic work of the Serenissima's official cartographer, Coronelli, particularly following the Morean War (Stouraiti 2002). Military strategy and imperial ambitions often manifested through the confessional power hierarchies. In this sense, elements of imperialism intertwined with confessional issues, economic factors with cultural aspects, graphic symbolism with textual narratives, and realistic themes with imaginary content.

Control of the Adriatic, along with the annexation of much of the north-eastern coast, was a key factor in establishing and maintaining the Venetian maritime empire. In this context, the graphics found in title cartouches and other auxiliary elements on nautical charts served as important visual tools for conveying political and symbolic messages. One notable cartographic depiction of Venetian Dalmatia created by Gaspar Tentivo and engraved by Francesco Felice Polanzani features a female figure adorned with the insignia of the Doge. She is seated next to the winged lion of St Mark, which serves as a powerful symbol of Venetian authority and protection (see Chap. 4, Fig. 4.1). The depiction of the glorious Republic of Venice (Serenissima Repubblica di Venezia), supported by one of the Evangelists, highlights the Church's protective role. It emphasizes Venice's status as the guardian and overseer of the Adriatic and its overseas possessions. Religious themes in cartographic representations are a consequence of the Venetian-Ottoman conflicts and an integral part of Venetian iconography. In addition to the winged lion of St Mark, the figure of the Virgin Mary is a significant icon. According to legend, the foundation of Venice occurred under her protection on March 25, 421, the day of the Annunciation. The Virgin Mary became even more prominent during the outbreak of the plague in 1630 (Frank 2019). Thus, the Adriatic landscape is cartographically represented with Venetian state symbols and illustrating values of Christian piety.

3.3 *Cultural-Ideological Rhetoric: Iconography of the Other on the Nautical Charts and Maps of the Adriatic*

Visual representations of one's own space, as well as the spaces of others, are reflected in the cartographic images that depict both the environment and its inhabitants. These images are created using various pictorial motifs found in the analysed material. Each individual cartographer's interpretation of geographical reality is unique, shaped by their personal imagination of space. This uniqueness is influenced by the author's religious, ethnic, political, or cultural identity (Mlinarić and Gregurović 2011; Krleža and Mlinarić 2022). Mapping space serves to reconceptualize and represent the identities of its inhabitants. Thus, maps are more than just records of spatial organization or contributions to geographical knowledge; they often carry layers of difference and "otherness" related to the depicted subjects. These subjects may differ from the cartographer in various ways, including political, cultural, ethnic, or religious affiliations, as well as factors such as age, gender, or economic status. The reconstruction of space in Western maps of the Adriatic illustrates that symbols used to identify the "other" are primarily employed to defend the cartographer's worldview and to challenge various ideologies. These symbols can range from religious identifiers (like Christian or Islamic) to secular ones (such as social or political markers) (Krleža and Mlinarić 2022).

An intriguing example of the depicted level of difference and "otherness" can be seen in a map from the Vallard Atlas, which is one of the sixteenth-century cartographic works produced in Dieppe, Normandy. This atlas illustrates the Adriatic Sea (Fig. 5.2) and features cartographic representations of the New World that maintain the foundational

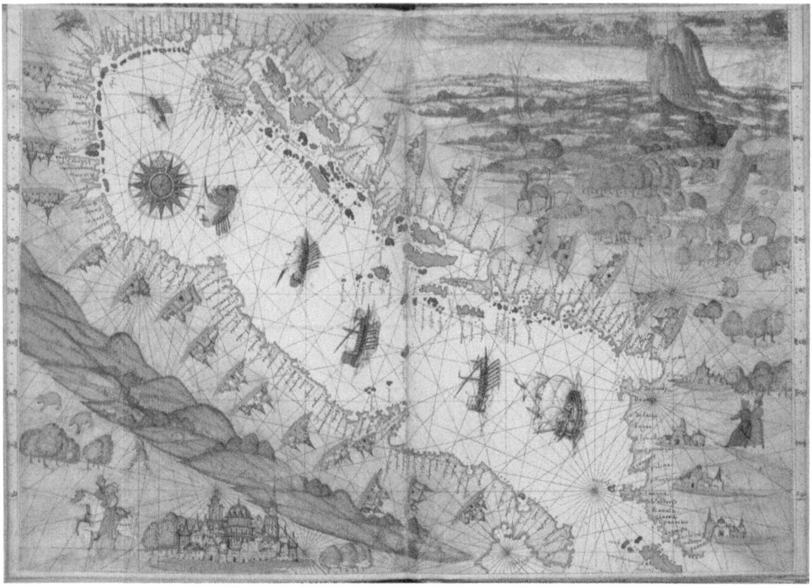

Fig. 5.2 Unknown author, Depiction of the Adriatic, Dieppe, 1547 (Source: Berkeley Digital Library, Henry E. Huntington Library and Art Gallery, San Marino, CA, HM 29, f. 14)

structure of nautical charts while also emphasizing more decorative depictions of unfamiliar lands. A defining feature of the maps in the atlas is the prominent use of motifs from the natural world, adorned with richly detailed depictions of flora and fauna. These embellishments were crafted to appeal to the perspectives and tastes of Western European buyers or patrons, with the aim of attracting a broader audience (Serchuk 2019). The map serves as both a representation of the terrain and a visualization of knowledge, requiring comprehensive information without any gaps or omissions. The richness of its content—regardless of whether it is fanciful or purely decorative—greatly enhances the professional credibility of the map, while also increasing its commercial value (Mitchell 1986).

The intricately designed and richly decorated nautical chart of the Adriatic by the unknown author from Dieppe, a French nautical cartographic centre in the sixteenth century, highlighted the political landscape, particularly the presence of the Ottomans in south-eastern Europe. This nautical chart of the Adriatic Sea is part of the so-called Vallard Atlas, named after Nicholas Vallard, the first owner of this 1547 publication. The cartographic depiction clearly distinguishes two zones: the familiar and the unfamiliar. The coastal areas—specifically the western and eastern coasts of the Adriatic—represent the known zone, enhanced by vedute and the naming of seaports and islands. This also applies to the Apennine Peninsula, where the largest veduta is located. The familiar coastal region of the Aegean is illustrated through three vedute and named ports. Additionally, the map features various ships, including merchant vessels and warships (galleys), highlighting the strategic significance of the Adriatic shipping route (Fig. 5.2).

In contrast to the known Adriatic coastland area, the further south-eastern European hinterland is portrayed as largely unexplored territory, characterized mainly by vast, uninhabited forests. The presence of the Ottomans is notably emphasized in this portrayal. An intriguing aspect of a nautical chart from Vallard's 1547 atlas is the selection of animals included in this area: the anonymous cartographer chose to illustrate exotic creatures such as elephants and camels instead of the native wildlife of south-eastern Europe. This choice exemplifies the exoticization of the distant "Other", as it features imaginary animals along the eastern Adriatic coast that do not actually exist. Additionally, this depiction enhances the exoticization of unfamiliar regions and their inhabitants by underscoring ethnic and cultural differences, particularly through the portrayal of local clothing compared to what was deemed "standard" in Western ethnographies.

The same cartographic circles from Dieppe influenced the production of the sixteenth-century nautical charts of the Adriatic from the so-called Dauphin Atlas, composed of maps of Europe, Africa, and America. One notable nautical chart of the Adriatic Sea (Fig. 5.3) features a golden fleur-de-lis to indicate the north and a silver crescent moon for the south; however, the chart is oriented with the south at the top.

Similar to the Vallard atlas and the accompanying nautical charts, this atlas was richly decorated to captivate the viewer's interest and ultimately gain the favour of the king. The nautical chart demonstrates the relationship between cartography and other contemporary pictorial (artistic) traditions of the time. From the humanist perspective, the concept of painting as a form of poetry served as the foundation for the development of the pastoral tradition, which drew themes from classical mythology. This notion became widely accepted and laid the groundwork for the pastoral landscape in art (Ruff 2015). The pastoral landscape tradition is balanced

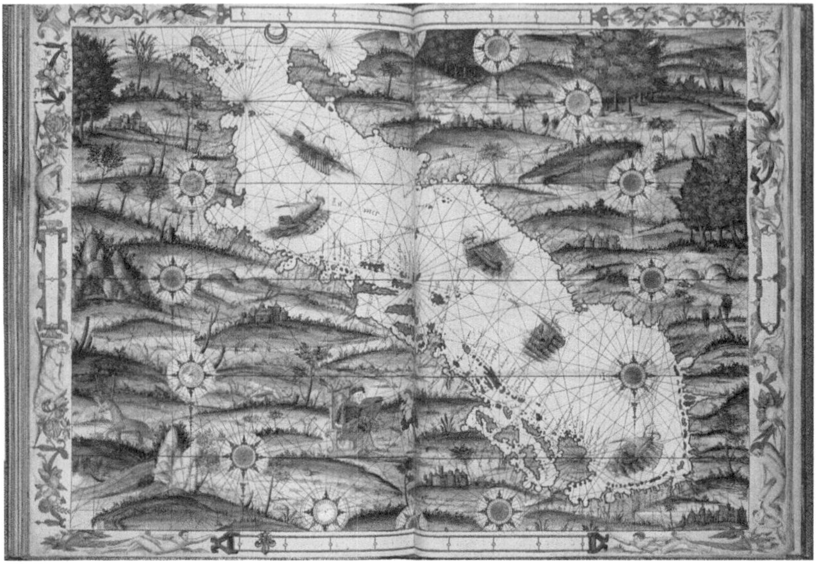

Fig. 5.3 Unknown author, nautical chart of the Adriatic Sea, in: Dauphin atlas, Dieppe, c. 1538–1546 (Source: Koninklijke Bibliotheek, Nationale bibliotheek, The Hague; 129 A 24, 29v-30r)

between classical bucolic poetry, Virgil's idealization of rural life, and the discoveries of the New World, along with advancements in technology and science. The decoration of the nautical charts should also be seen in this context, as it not only enhanced the atlases and added to their visual appeal but also assisted viewers in understanding the maps and nautical charts. In this way, cartographers actively experimented with format and decoration to integrate the new image of the world into the existing framework for disseminating knowledge (Serchuk 2019). For example, a nautical chart by an unknown author illustrated a well-known area—the Adriatic Sea. However, this region was somewhat distant for both customers and cartographers, particularly those in Dieppe. This unfamiliarity is evident in the incorrect perception of the Adriatic Sea's location, which was depicted further east and mistakenly referred to as *La Mer Rouge* (Red Sea) (Faričić et al. 2023b). The Adriatic was depicted in a humanist style reminiscent of an idealized literary Arcadia, which was visually translated into a cartographic Arcadia. The illustration featured a vibrant landscape that combined both natural and cultural elements. The political climate and the presence of the Ottomans in the eastern Adriatic hinterland certainly influenced the cartographic perception of this region as exotic, notable for its unique fauna and flora. In cartographic representations, the Ottoman territory was typically marked with the political symbol of the crescent moon. Culturally, this presence was often depicted through decorative elements featuring exotic vegetation and animals, reflecting the style typical of Arabic Proto-Ottoman medieval miniature art (Mlinarić and Miletić Drder 2017). The unknown author of the Dauphin Atlas portrayed both Adriatic coasts as a familiar zone and harmonized their content. On these coasts, the author depicted urban structures without establishing any hierarchy among them. In addition to the exotic cartographic representation of Arcadia, this nautical chart exemplifies the concept of cartographic syncretism, which is achieved through the fusion of different rhetorical styles (see also Fig. 5.2). The small image (miniature) of two human figures—one crowned and the other naked and ashamed—resembles the biblical iconography of Adam's expulsion from the Garden of Eden. This motif further illustrates the intricate combination of rhetoric applied to the territories of the eastern Adriatic on the same map.

3.4 *Rhetoric of Hierarchy: Selected, Suppressed, or Misrepresented Content*

The selection and organization of content on early modern nautical charts allowed chart authors to omit areas they did not know well, as well as regions that did not align with their interests and that they deliberately chose to conceal. A common example of this practice was the frequent omission of the Ottoman presence or borders, which will be discussed in further detail later. Additionally, cartographers often included decorative elements to fill in these gaps, attempting to disguise their lack of geographical knowledge about certain areas and enhance the commercial value of their products. The development of cartography during the early modern period was influenced by various cultural and geopolitical processes, characterized by a contradictory yet harmonious interplay between traditional and modern elements.

This mutual contribution also enables the cartographic conceptualization of geographical, historiographical, and philosophical ideas such as continuity and discontinuity, duration and change, as well as civilizational, cultural, economic and social exclusion, oppression, and penetration in the Adriatic region, which served as a European-Mediterranean intersection from the sixteenth to the nineteenth century. Cartographic production not only relied on the individual author's personal drawing skills and geographical knowledge but also reflected ideological and political attitudes and perspectives, advancements in technology, and scientific progress.

Each cartographic tradition highlighted its own mapping knowledge as a form of mediated truth, and propaganda. By shaping a "biased" visualization of space, maps indirectly conveyed the political ideas and interests of the cartographic tradition that created them. Therefore, as a means of communication among various actors, maps also supported the state interests of the Republic of Venice, serving as an important medium for producing and showcasing "Venetian knowledge"about space (Stouraiti 2016). This phenomenon was evident in both private and public spheres. Venetians fondly used cartographic representations as decorative elements in their homes and public spaces. The cartographic choices of the elites did not primarily feature Venice itself but highlighted the Republic's territorial accomplishments. Subtle messages that celebrated the Republic's territorial and commercial policies were further reinforced when family members or their ancestors played a role in expanding the Venetian empire (Carlton 2012). In this way, maps served as a means to promote awareness of unknown territories and expansionist ambitions. However, the fear of the unknown also significantly influenced how lesser-known lands were portrayed. The cartographers' ignorance or lack of good knowledge of distant lands has already been acknowledged, particularly in the works produced by foreigners regarding our regions. In this context, we can observe that terrifying sea monsters are depicted in the Adriatic, which contrasts with the extraordinarily detailed and richly decorated nautical charts created by Venetian experts in this area. These monsters are typically found only as generic symbols representing distant and unknown lands on nautical charts made by cartographers from other Western cartographic traditions, such as the Dutch or French (Fig. 5.4). The presence of sea monsters like oversized whales, in the cartographic works of Barents and Roux, symbolizes the unknown and reflects the mysteries associated with the depths of the sea.

The era of overseas expansion brought influences from classical and Islamic Mediterranean traditions into cartographic representations. According to the ancient Greek concept of climatic zones, the ecumene refers to the central belt surrounding the Mediterranean, which was

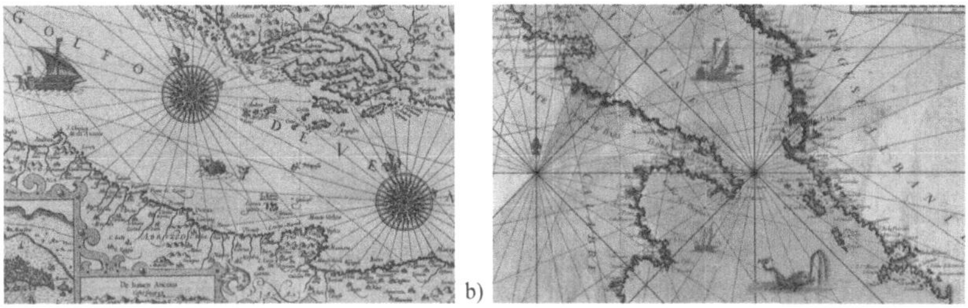

a) b)

Fig. 5.4 Depiction of the sea monsters on the nautical charts of the Adriatic Sea: (**a**) segment of Willem Barents' nautical chart, 159 (Source: The Barry Lawrence Ruderman Map Collection, Stanford University Libraries, Tooley, A-D:85, 268; K-P:10); (**b**) segment of Joseph Roux's nautical chart, Marseille, 1764 (Source: National and University Library, Map and Atlas Collection, Zagreb, S-JZ-XVIII-116)

deemed suitable for civilized life. Everything beyond the ecumene was considered an unknown space inhabited by monsters, a concept that dates back to Herodotus and Pliny and was depicted in medieval bestiaries. Typically, these monsters were portrayed as hybrid beings dwelling in wild spaces, existing somewhere between human and animal forms. Consequently, exotic ornaments were added to early medieval nautical charts in regions that were less familiar or depicted as "emptied" spaces. This was done to enhance marketing appeal and cater to the public's fascination with imaginative and decorative cartographic content. The geostrategic position of the Adriatic Sea places it at the intersection of various European imperial states and two major religious systems: Christianity and Islam. This unique position is reflected, among other things, in cartographic patterns that extend beyond the primary purpose of nautical charts, which serve as critical expressions and representations of spatiality related to the sea and geographical realities in the Mediterranean. Symbolic elements found in early modern cartographic representations of the Adriatic highlight its cultural and religious connections to the European maritime landscape, specifically within the Christian cultural sphere. These nautical charts reveal broader imagological and communicative capacities of cartographic media, which were intended for a much wider (or different) audience than just sailors. The most prominent maps that illustrate the expansion of Venetian influence and depict the north-eastern Adriatic coast during the early modern period are primarily of Venetian provenance. As a key political and commercial centre with an economy reliant on maritime trade, Venice required accurate cartographic representations of its trading territories on land (*terraferma*) and at sea (*stato da mar*) (Juran et al. 2020).

The shipping route along the north-eastern Adriatic coast was crucial for transportation, serving as a key connection between Europe and the Mediterranean. As a result, Venice was keenly interested in accurately understanding the geographical features of this area, which naturally led to the development of increasingly precise cartographic representations (Mirošević and Faričić 2022). There was also a deliberate manipulation through the omission or misrepresentation of known geographical information, particularly evident in Venetian depictions of borders with the Ottomans over land that once belonged to them. The Venetians were reluctant to acknowledge these new political boundaries and the reduction of their territory, as it reflected their military and strategic inferiority during the Ottoman expansion. Consequently, they strategically and consistently avoided depicting these borders on nautical charts of Venetian Dalmatia. In contrast, they swiftly mapped their newly conquered territorial possessions as quickly as possible, marking them as new acquisitions. A similar selective practice was observable among numerous Western cartographers in the early modern period. They often used toponymy and even graphics to emphasize the ancient identity of the Adriatic coasts, the importance of the proto-Ottoman heritage, memory of medieval Christian kingdoms, and, more broadly, the Christian identity and affiliation of the Croatian lands until the eighteenth century. By doing so, cartographers neglected, ignored, or refused to document the Ottoman border neighbourhood as an essential aspect of the military-strategic landscape at the time (Mlinarić and Miletić Drder 2017; Krleža and Mlinarić 2022).

Opposing the same principle, Joan Martines' nautical chart of the eastern Mediterranean (including the Adriatic) reflects the political landscape of the time (Fig. 5.5). Three vedute, adorned with a crescent moon, illustrate the territories of south-eastern and eastern Europe that were under Ottoman rule. In contrast, two vedute bearing the heraldic arms of the ruling governments of the cities in the area represent the Adriatic hinterland. The nautical chart prominently features Venice, marked by the flag of the winged lion, a powerful emblem of its maritime dominance and influence. Complementing this imagery is the bell tower of St Mark's Basilica (St Mark's Campanile), which represents both political significance and economic prosperity, reflecting the city's wealth and power.

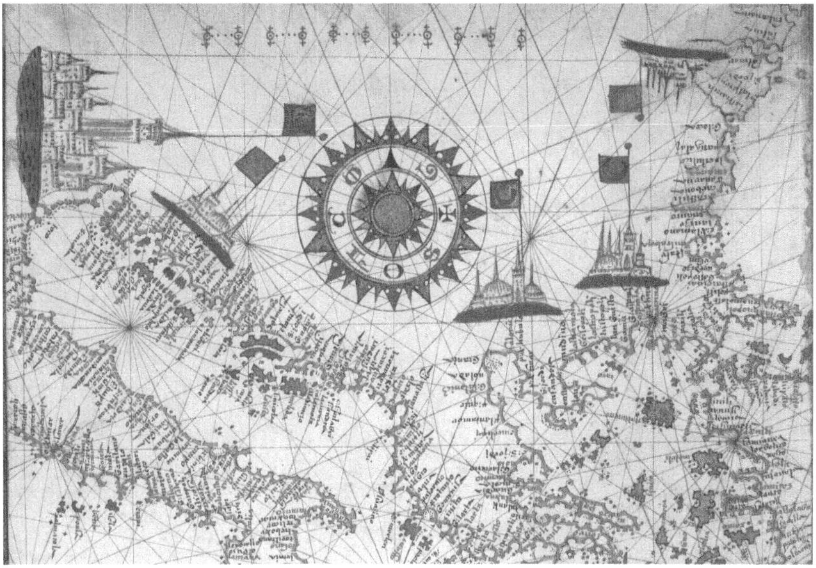

Fig. 5.5 The depiction of the Adriatic Sea on Joan Martines' nautical chart of the eastern Mediterranean, c. 1578 (Source: The Huntington Library, Art Museum, and Botanical Gardens, Library Collections, Maps and Atlases, Portolans, San Marino, CA, USA, mssHM 33)

Beyond its symbolic importance, the bell tower also served as a practical navigational aid and an imposing landmark, towering over the surrounding landscape. It likely left a lasting impression on every sailor who saw it.

The depiction of Venice dominates the map, marked by the flag of the winged lion, which symbolizes its maritime power and influence. Additionally, the Ottoman cities are marked with crescent moon signs, with no particular hierarchy among them. In contrast, the European cities are individually labelled with their heraldic identities, suggesting a higher importance. Demarcation for the city of Venice is identical to the identification of the state authority.

However, contrary to the principle of omitting information, there is also a tendency to overemphasize important ideas or exaggerate political facts. A notable example of this is the seventeenth-century nautical chart of the Adriatic created by Niccolò Guidalotti, which was one of the most artistic charts from the Venetian provinces (Fig. 5.6). This nautical chart

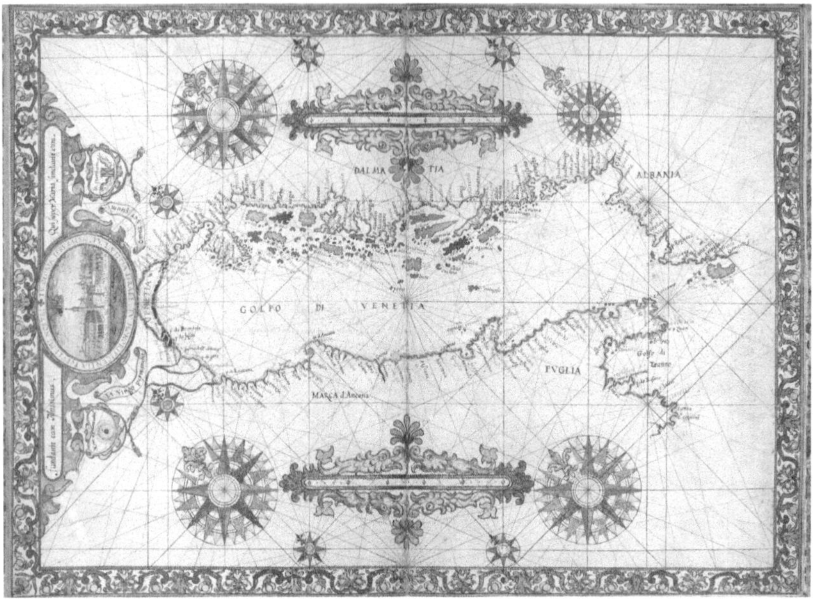

Fig. 5.6 Niccolò Guidalotti, nautical chart of the Adriatic Sea, in: Nautical Atlas, Venice 1646 (Source: Biblioteca Nazionale Marciana, It. IV, 10=5062; f. 2)

exemplifies the concept of communication hierarchy in maps by highlighting the supremacy and significance of the Venetian capital city. It is part of the Nautical Atlas, and Guidalotti dedicated it to the Venetian ambassador Giovanni Soranzo (Debby 2019).

The geographically empty parts of the nautical chart field were adorned with decorative motifs divided into three sections. At the edge, an oval veduta of Venice is prominently featured, richly embellished with two medallions. Both medallions depict allegorical figures and messages, one with a winged hourglass symbolizing eternity, accompanied by the inscription in Latin: *in eternum* ("forever"). The other portrays the Ptolemaic view of the universe, with Earth depicted and the phrase in Latin *non commo uebit* ("it will not be easy"). The most striking feature is an oval veduta of Venice, framed by inscriptions that honour the city and its origins: *Fundavit eam Altissimus, qui super Maria Fundavit eam and Tu Gloria Ierusalem, Tu Laetitia Israel, Tu Honorificentia Populi dei Nostri* ("The Most High founded it, who founded it on Mary, and You are the

Glory of Jerusalem, You are the Joy of Israel, You are the Honorificence of the People of Our God"). The perspective of the vedute is from the seaward side of Venice, revealing the lagoon and leading out to the Piazzetta di San Marco, where monumental structures like the Doge's Palace and the Palace of the Sansoviniana Library (now Biblioteca Nazionale Marciana) can be seen. The importance of that impression lies in the fact that the graphic is highly detailed while accurately depicting the key Venetian buildings and monuments. Particularly remarkable depictions are the symbols of the two patron saints of Venice (St Theodore, the patron saint of the city before St Mark, and the winged lion, known as the Lion of Venice), both represented on large columns. Other prominent buildings included in the graphic are the Basilica of San Marco, the Campanile, and the Clock Tower of San Marco, all adorned with gilded statues on their rooftops (Debby 2019). The nautical chart also reflects Venetian heraldic identities. Additionally, there are two almost identical decorative elements at the top and bottom of the chart: compass roses and linear scale without units of measurement. Venetian cartographers distinguished their work by incorporating elegant dolphin motifs into their maps, setting themselves apart from other traditions that often featured fearsome sea monsters. Instead of depicting these mythical creatures, the Venetians focused on aesthetic designs that adorned the empty spaces of their charts. The centrepiece of these nautical charts was a powerful symbol of Venetian authority, particularly evident in Guidalotto's atlas, which was dedicated to the Venetian ambassador in Constantinople. This was a deliberate choice to showcase the Republic of Venice's heraldic emblem, complemented by a detailed city view that asserted Venetian dominance over the Adriatic Sea, known as the *Golfo di Venezia*. The artistry and intention behind these nautical charts not only reflect the cultural pride of Venice but also serve as a testament to its unparalleled maritime supremacy. As far as the chart decoration is concerned, the style of the atlas and the accompanying nautical charts reflect the direct influence of the well-known Catalan Oliva family of cartographers. The members of this family were very active in cities such as Venice, Messina, Naples, Livorno, and Marseille. The decorative elements in their atlases were designed to captivate wealthy patrons (Debby 2019). This focused effort to visually enhance the importance and relevance of Venice exemplifies a clear technique of power communication.

4 Conclusion

The decorative cartographic motifs found on early modern nautical charts and maps of the Adriatic serve purposes beyond simple geographical illustration and spatial representation. These embellishments highlight the unique position of the Adriatic at the intersection of various European political, cultural, and religious systems, each representing its own interests and territorial claims. Apart from their intended use and the intended audience, the artistic and iconographic elements showcase the cartographers' geocartographic expertise, as well as their skills in engraving, painting, and overall artistry.

A comparative, interdisciplinary analysis of selected nautical charts of the Adriatic, which were created for different purposes and at different scales, reveals several dominant forms of communication through decorative imagery. These iconographically designed geocartographic elements can be divided into three interconnected communication concepts: the rhetoric of decorative elements, the hierarchy of representational content, and the suppression or misrepresentation of accurate geocartographic information. While these modes of communication rarely function independently, together they illuminate the multi-layered messages conveyed by the maps. In particular, decorative elements often reflect geostrategic ambitions, such as the Venetian ideologies of expansion. Another level of rhetoric arises from cultural-ideological constructs that portray "others" as different from the majority population in the Adriatic, frequently implying notions of inferiority or exoticism. The hierarchy of the selected content also emphasizes the prejudices inherent in the cartography. Suppressed or inaccurately depicted elements reveal the biases present in cartography. Elements that are suppressed or inaccurately represented reveal the intentions, limitations, or deliberate omissions of the cartographer. These choices reflect inherited perceptions, established conventions, and evolving attitudes towards the landscape and its inhabitants. The decorative graphic motifs on maps, particularly those produced in European workshops, often reinforced older ideological frameworks, including the Christian worldview.

Venetian cartography, exemplified by the work of Vincenzo Maria Coronelli, played a leading role in this context. Venice utilized cartographic representation as a tool of state authority to project its political and religious mission of sovereignty onto newly acquired territories. Coronelli's maps celebrated Venetian military campaigns and territorial conquests, depicting the Republic as a dominant power with a divinely ordained purpose. This imperial narrative also influenced the representation of the local population, portraying them as subordinate or "inferior" within the Venetian hierarchy. Other European cartographers often shared these views, further reinforcing a common discourse of supremacy.

A particular aspect of these maps is the identification and representation of the unknown and the exotic. This was often achieved through the use of fantastical or mythological images, such as monstrous sea creatures like large whales and exotic animals. These motifs served a dual purpose: they symbolized the untamed or mysterious regions of the east Adriatic while also fulfilling an aesthetic function by filling empty spaces in lesser-known geographical areas on the map. Such images reinforced recently produced mental maps and emphasized the cultural demarcation between the familiar, "civilized" space of the Western Adriatic coastline and the "wild" or "foreign" territories of the eastern Adriatic hinterland.

Ultimately, these maps served as powerful tools for conveying geopolitical, cultural, and religious messages. By distinguishing "our"space from the space of the "other", they reinforced territorial, ideological, and cultural boundaries while reflecting the broader aspirations and fears of their creators and patrons.

References

Blaut JM, Stea D, Spencer C, Blades M (2003) Mapping as a cultural and cognitive universal. Ann Assoc Am Geographers 93(1):165–185. https://doi.org/10.1111/1467-8306.93111

Brunn SD, Dodge M (2017) What is where? The role of map representations and mapping practices in advancing scholarship. In: Brunn SD, Dodge M (eds) Mapping across academia, 1–22. https://doi.org/10.1007/978-94-024-1011-2_1

Carlton G (2012) Making an impression: the display of maps in sixteenth-century Venetian Homes. Imago Mundi 64(1):28–40. https://doi.org/10.1080/03085694.2012.621394

Cartwright W (2009) Art and cartographic communication. In: Cartwright W, Gartner G, Lehn A (eds) Cartography and art. Springer-Verlag, pp 9–22. https://doi.org/10.1007/978-3-540-68569-2_2

Cosgrove D (2008) Geography and vision: seeing, imagining and representing the world. I. B. Tauris, London and New York

Debby NB-A (2019) Fra Niccolò Guidalotto's City View, Nautical Atlas and Book of Memories: cartography and propaganda between Venice and Constantinople.

In: Baumgärtner I, Debby NB-A, Kogman-Appel K (eds) Maps and travel in the Middle Ages and the Early Modern Period: knowledge, imagination, and visual culture. De Gruyter, Berlin, Boston, pp 9–22. https://doi.org/10.1515/9783110588774-014

Faričić J (2007) Geographical names on 16th and 17th century maps of Croatia. Kartografija i geoinformacije. 6(spec.): 149–179.

Faričić J, Kljajić I, Mirošević L, Mlinarić D (2023a) Symbolism of compass roses on Early Modern Nautical Charts of the Adriatic Sea. KN – J Cartography Geographic Inf73(1):19–37. https://doi.org/10.1007/s42489-022-00129-z

Faričić J, Selva O, Umek D (2023b) Geographical names of the Adriatic Sea on medieval and early-modern maps and nautical charts. J Hist Geography 82:68–80. https://doi.org/10.1016/j.jhg.2023.09.007

Frank M (2019) Representing the Republic in Seventeenth-Century Venice. Radovi Instituta povijesti i umjetnosti 43:113–122. https://doi.org/10.31664/ripu.2019.43.09

Duzer C (2021) Colonialism in the Cartouche: imagery and power in early modern maps. Stud Class Trad 9(2):90–130. https://doi.org/10.20396/figura.v9i2.15620

Harley JB (1988a) Silences and secrecy: the hidden agenda of cartography in early modern Europe. Imago Mundi 40(1):57–76. https://doi.org/10.1080/03085698808592639

Harley JB (1988b) Maps, knowledge, and power. In: Cosgrove D, Daniels S (eds) The iconography of landscape: essays on the symbolic representation, design and use of past environments. Cambridge University Press, Cambridge, New York, pp 277–312. https://doi.org/10.1080/03085698808592639

Harley JB (1992) Deconstructing the Map. In: Barnes T, Duncan J (eds) Writing worlds. Discourse, text and metaphor in the representation landscape. Routledge, London and New York, pp 231–247

Ivetic E (2019) Storia dell'Adriatico: un mare e la sua civiltà. Società editrice il Mulino

Juran K, Barzman K, Faričić J (2020) Kartografija u službi Mletačke države: karta sjeverne i srednje Dalmacije nepoznatog autora s početka 16. stoljeća. Geoadria 24(2):93–139. https://doi.org/10.15291/geoadria.2885

Krleža P, Mlinarić D (2022) Visual images on Medieval and Early Modern world maps: iconography of the known and the unknown. Cartography Geoinfor 21(38):26–51. https://doi.org/10.32909/kg.21.38.2

Mayhew T (2008) Dalmatia between Ottoman and Venetian Rule – Contado di Zara 1645–1718. Roma

Mirošević L, Faričić J (2022) The Island of Korčula on early modern maps. Cartography Geoinf 21(spec. issue):99–125. https://doi.org/10.32909/kg.21.si.8

Mitchell TWJ (1986) Iconology: image, text, ideology. University of Chicago Press

Mlinarić D, Miletić Drder M (2017) Zbirka Novak: Mappae Croaticae u Zbirci zemljovida i atlasa NSK. Nacionalna i sveučilišna knjižnica u Zagrebu
Mlinarić D, Gregurović S (2011) Kartografska vizualizacija i slika Drugoga na primjeru višestruko graničnih prostora. Migracijske i etničke teme 27(3):345–373. https://hrcak.srce.hr/78010
Monmonier M (2018) How to lie with maps, 3rd edn. University of Chicago Press
Panofsky E (1970) Meaning in the visual arts. Penguin
Ribeiro DM, Caquard S (2018) Cartography and art. In: Wilson JP (ed) The Geographic information science & technology body of knowledge. UCGIS. https://doi.org/10.22224/gistbok/2018.1.4
Rees R (1980) Historical links between cartography and art. Geographical Rev 70(1):61–78
Robinson AR (1982) Early thematic mapping in the history of cartography. University of Chicago Press, Chigaco and London
Ruff RA (2015) Venice and the pastoral landscape. In: Arcadian visions, pastoral influences on poetry, painting and the design of landscape. Windgather Press, Oxbow Books, pp 35–43. http://www.jstor.org/stable/j.ctt19704rw.8
Serchuk C (2019) Around the world: borders and frames in two sixteenth-century Norman map books. In: Baumgärtner I, Ben-Aryeh Debby N, Kogman-Appel K (eds) Maps and travel in the Middle Ages and the Early Modern Period. Knowledge, imagination, and visual culture. De Gruyter, Berlin, pp 189–218. https://doi.org/10.1515/9783110588774
Stouraiti A (2002) Propaganda figurata: geometrie di dominio e ideologie veneziane nelle carte di Vincenzo Coronelli. Studi veneziani 44:129–155
Stouraiti A (2016) Printing empire: visual culture and the imperial archive in seventeenth-century Venice. Hist J 59(3):635–668. https://doi.org/10.1017/S0018246X1500031X
Šakaja L (2015) Uvod u kulturnu geografiju. Leykam international d.o.o, Zagreb
Wintle M (1999) Renaissance maps and the construction of the idea of Europe. J Hist Geography 25(2):137–165. https://doi.org/10.1006/jhge.1999.0111

CHAPTER 6

Religious Symbolism on Old Nautical Charts of the Adriatic Sea

Josip Faričić

1 Introduction

During the Middle Ages and early modern period, religion profoundly influenced European society, permeating nearly every aspect of life. Its impact extended beyond traditional religious practices, shaping perceptions of spatial reality—from cosmological ideas to governance at the micro-regional level. Consequently, sacred elements became integral to the cultural landscape, appearing iconographically in literature, art, and cartography. This influence is particularly evident in maps and charts that sought to represent space and spatial relationships. Spatial depictions during this period were intricate composites of the real and imagined worlds, making it challenging to distinguish between the two. Mental ambiguity regarding the concept of space was common, especially when space was viewed as a social construct rather than a geometrically defined portion of the known world (Paasi 2000; Forsberg 2003; Summa 2017). Real space was not always physically defined—visible or measurable—and often encompassed both material and spiritual dimensions of human

J. Faričić (✉)
Department of Geography, University of Zadar, Zadar, Croatia
e-mail: jfaricic@unizd.hr

J. Faričić, T. Marelić (eds.), *Early Modern Nautical Charts of the Adriatic Sea*, https://doi.org/10.1007/978-981-96-8046-7_6

relationships with places. These relationships varied depending on individual and communal perceptions, experiences (Buttimer 2015), existing knowledge, and even prejudices, aspirations, and expectations about a given space. Imagined space, however, was not merely the product of spontaneous imagination or fantasy. When imagination was embedded in mythological images, it often corresponded to real spatial facts, processes, or relationships. For instance, the adventures of the Argonauts in Apollonius's *Argonautica* reflect historical connections between the ancient Greek world, the Black Sea, and peripheral Mediterranean regions (Meyer 2008). These mythical narratives inspired the Venetian cartographer Vincenzo Maria Coronelli, who named the geographical society he founded in 1684 the *Accademia Cosmografica degli Argonauti* (Milanesi 2016). Furthermore, the myth of the Argonauts left tangible traces in real-world geography. For example, the ancient Greek name of the Kvarner Islands, *Absyrtides*, was derived from *Absyrtus*, the mythical brother of Medea (Katičić 1968; Kozličić 1990). This figure from Greek mythology is still commemorated today in the name of the settlement of Osor on the island of Cres in the Adriatic Sea. The etymology of Osor traces back to the Greek name *Absyrtides*, through Latin (*Apsorus*), Italian (*Ossero*), and eventually Croatian (*Osor*). This enduring legacy illustrates the complex interplay between myth, imagination, and geography throughout history.

The dichotomy and interplay between real and imagined spaces parallel the relationship between profane and sacred spaces, with perception serving as a crucial factor in both (Jackson and Henrie 1983). Sacred space is not merely a segment of the broader spatial realm tied to religious practices. Its full significance emerges through its interconnection with profane space, enriching everyday life for communities of believers. Religiosity extends beyond formal worship in sacred spaces, encompassing broader spatial dimensions, much like spirituality influences both individuals and communities across various facets of life and activity (Sopher 1967; Park 1994; Holloway and Valins 2002; Stump 2008). Religious motifs often manifest as enduring symbols with spiritual imprints on material culture. In communication, these motifs signify belonging and devotion, transcending distinctions between religions. This dynamic found a distinct expression in historical cartographic practices.

Cartography, a complex process of graphically representing space, employs artistic techniques and patterns to convey ideas, perspectives, experiences, and emotions about both real and supernatural (religious, mythological, fantastical) phenomena through a two-dimensional visual language. This methodology, heavily influenced by fine arts, led cartographers to navigate between science and art for centuries (Ribeiro and Caquard 2018). This interplay contributes to the ongoing debate about the nature of cartography, emphasizing the inseparable intertwining of art and science (Karssen 1980; Krygier 1995; Cartwright 2009). Iconographic patterns rich in symbolism, commonly used in fine art, were similarly incorporated into cartography. Old maps, including nautical charts, were imbued with symbolic content that extended beyond simple geographic encoding, offering layers of meaning and interpretation. During the late Middle Ages, they reflected the artistic styles found in illuminated manuscripts, while in the early modern period, they aligned with the elaborate artistic expressions of the Renaissance and Baroque eras.

While primarily designed for navigation, focusing on coastlines, ports, anchorages, and hazards (Campbell 1987; Astengo 2007), nautical charts also served a broader audience. Nautical charts were not solely utilitarian tools for sailors navigating specific tasks. They were also created for shipowners, scholars, and members of secular and religious elites—individuals invested in the politically and economically dynamic marine and coastal regions of the time. For these audiences, utilitarian content was often enriched with aesthetic elements, such as intricate decorations (Sheehan, 2012). These embellishments elevated nautical charts into artefacts of beauty, akin to illuminated manuscripts, blending practical utility with artistic elegance (Gaspar and Leitão 2018, 2019; Schilder and Egmond 2007). In creating their nautical charts, cartographers sought to position decorative elements, including religious motifs, in areas of the chart that would not obscure important geographical details in coastal regions. They often did this on the so-called neck of the nautical chart (Astengo 2007). An example of such a nautical chart is the 1615 map of the Mediterranean Sea created by Sebastiano Condina in Naples (Fig. 6.1). In the hinterlands of Mediterranean regions, he illustrated animals, cities, and compass roses, while the western margin featured a depiction of the Virgin Mary with the infant Jesus. This iconographic element is associated with the veneration of St Mary as the protector of sailors, who honoured her with the title *Stella Maris*—Star of the Sea (Canney 1937; Santoro 2011). Alongside St Mary, sailors also sought the intercession of St Nicholas. This devotion was motivated by the general religiosity of sailors and the frequent uncertainties and dangers they faced, which, given the technical features and manoeuvring capabilities of their ships and the available knowledge of terrestrial and astronomical navigation, often exceeded their skills and

Fig. 6.1 Sebastiano Condina, nautical chart of the Mediterranean Sea, Naples, 1615 (Source: *Biblioteca Nazionale Marciana*, It. IV, 505 (=10036), t. 1)

expertise (Gambin 2014). While Condina's nautical chart reflects the imprint of Marian devotion in cartography, the spiritual attachment to St Mary, one of the most prominent saints across all Christian denominations, was expressed by sailors in numerous other ways. These included the establishment of maritime pilgrimage centres (Katić and Vedriš 2023) and the offering of votive gifts featuring images of ships that included depictions of the Blessed Virgin Mary, which were displayed in churches (Kisić 1999; Timmermann 2021).

One of the maritime areas depicted on old nautical charts is the Adriatic Sea, a part of the Mediterranean nestled between the Alps, Apennines, and Dinarides. For centuries, it has served as a vital link between Europe, Africa, and Asia (Chaline et al. 2001). The Adriatic, described by Egidio Ivetic (2019) as the "Mediterranean within the Mediterranean," has long been a dynamic hub of cultural exchange and technological advancement. It has been divided among several political entities and located at the crossroads of Western and Eastern Christianity (Farinelli 1999; Ivetic 2019). Following the Ottoman conquests, it also became a contact zone between Christianity and Islam. These factors led to the incorporation of religious symbolism on nautical charts, serving not only as decorative elements but also as tools for identifying and distinguishing specific Adriatic regions.

2 Research Methodology

The Adriatic Sea was depicted on old nautical charts as early as the thirteenth century. It appeared on small-scale charts of the entire or central Mediterranean and on larger-scale maps where it occupied the entire field, which can be classified as nautical charts of the Adriatic. In addition to these charts, Adriatic ports and other smaller spatial units were also depicted on nautical charts, though rarely as standalone cartographic products. More often, they appeared as part of *portolani* (navigational manuals) and *isolarii* (comprehensive publications on islands). Through research into scientific literature and the collections of numerous European and American archives, museums, libraries, and private collections approximately one hundred nautical charts of the Adriatic have been identified, along with other types of maps sharing similar geographic content with nautical charts (Faričić et al. 2023b). Most of these charts were produced in cartographic centres around the Mediterranean, predominantly in Venice, Genoa, and Naples, with very few known examples of Ottoman origin.

Nautical charts of the Adriatic Sea were not equally rich in decorative elements, and not all of them included religious themes. However, from the Middle Ages through the early modern period, until the institutionalization of cartography and the geodetic and hydrographic surveys that preceded the creation of the first modern nautical charts in the early nineteenth century, it is evident that religious motifs (drawings and symbols) were present on numerous maps. These motifs appeared either explicitly or indirectly, often through heraldic elements influenced by iconographic patterns of religious significance.

The study did not delve into the differences in graphical techniques used to depict religious themes, or whether the nautical charts and similar maps were handcrafted or printed. In analysing nautical charts with religious symbolism, the method of deconstruction was applied, originally developed by John Brian Harley (1989) and further elaborated by Denis Cosgrove (1992, 1999) and other scholars (Edney 2005). Based on Harley's premise that it is crucial to determine why something is depicted or omitted on a map, rather than merely how and where it is depicted (Harley 1989), as well as Cosgrove's argument that maps can represent material and immaterial, existing and desired, as well as experienced, remembered, or projected elements of space (Cosgrove 1999), the study adopted a context-sensitive approach. Additionally, the idea that maps lack a fixed ontological status and must be analysed within their appropriate context and in relation to other methods of spatial representation (Kitchin and Dodge 2007) helped shape the research goal.

The primary aim was to identify the key functions of religious content on nautical charts of the Adriatic. A major research challenge in this context is that it is difficult to ascertain the intentions of the chartmaker, the commissioner, or anyone who influenced its artistic design. One can only presume that the religious motifs were related to the spirituality of these individuals, or that they were part of established iconographic patterns from the visual arts, transferred onto the charts without retaining the original semantics of those decorative elements. Consequently, the description of individual visual elements was deliberately minimized.

The difficulty in interpreting religious symbolism stems from the limited information available about the religious inclinations of the mapmakers. For instance, while it is reasonable to assume that V. M. Coronelli's spirituality significantly influenced his worldview and, consequently, the design of his maps—given that he was a priest and, at one point, the head of the *Order of Friars Minor Conventual* (Milanesi 2016)—the same cannot be definitively stated for many other cartographers. These individuals, often geographers, sailors, or professionals from other primary fields, leave us uncertain about whether the religious elements on their maps and charts reflect their personal spiritual beliefs or simply the general characteristics of an era in which religion permeated many aspects of society.

3 Religious Symbols on the Old Nautical Charts

Religious symbols on the old nautical charts of the Adriatic Sea are primarily rooted in Christianity but occasionally also contain Islamic motifs and reflect the historical coexistence and occasional conflict between the two religions in the Mediterranean.

3.1 Religious Symbols on Compass Roses

Among the earliest religious symbols used in nautical cartography are the fleur-de-lis, the cross, and the crescent moon, often found on compass roses. The fleur-de-lis, a stylized depiction of a lily, began to mark north on compass roses at the end of the fifteenth century. The first recorded use of this symbol to denote north on a nautical chart was by Jorge de Aguiar in 1492 (Campbell 1987). Beyond its practical purpose, the fleur-de-lis carried rich religious symbolism, representing the Blessed Virgin Mary, the mother of Jesus Christ, and the Holy Trinity. It also held significance as a heraldic emblem, prominently associated with nobility and ruling houses such as the French Valois and Bourbon dynasties, the related Anjou dynasty, and various English royal houses, including the current one (Feuillet 2007; Caldwell 2014). It is worth noting that the North (Polar) Star, a critical celestial reference for navigation on clear nights, was also referred to as *Stella Maris*—a title linked to the Blessed Virgin Mary (Vagnon 2021). On nautical charts of the Adriatic Sea, however, the fleur-de-lis did not consistently appear to mark north until the seventeenth century, with only sporadic exceptions during the sixteenth century (Fig. 6.2). It is noteworthy that the lily flower was used to indicate north on compass roses by the Ottoman sailor and cartographer Muhiddin Pîrî Reis (Faričić et al. 2023a). On his 1526 map of the Adriatic Sea, as well as on detailed charts of smaller maritime areas around Adriatic ports and islands, Pîrî Reis seems to have employed such a north marker likely without its

Fig.6. 2 Compass roses with the fleur-de-lis marking north and the sign of the cross marking east on a nautical chart of the Adriatic Sea: (**a**) chart made by Diogo Homem (Venice, 1559) (Source: *Bibliothèque nationale de France, département Cartes et plans*, Paris; CPL GE DD-2003 (RES)); (**b**) chart made by Vincenzo Demetrio Volčić (Naples, 1593.) (Source: *National Library of Finland*, Maps, The Nordenskiöld Map Collection, Helsinki; N-Kt-103c)

original semantic content, that is, without the Christian connotations of this heraldic symbol. Since only a limited number of Ottoman maps of the Adriatic have been preserved, no pattern can be definitively established; Pîrî Reis's use of this symbol appears to be an isolated case from which no definitive conclusions can be drawn, apart from the strong possibility that this Ottoman Muslim cartographer was influenced by the artistic conventions of European Christian cartographers (Soucek 1992; Goodrich 2004). The clash between the Ottoman Empire and European Christian powers in the Mediterranean during the early modern period did not prevent interactions among sailors and cartographers, thus continuing the cartographic dialogue between Christian and Islamic authors that had been established as early as the twelfth century in the works of Muhammad al-Idrisi (Maqbul Ahmad 1992).

In addition to the specific graphic marking for north, east was often represented symbolically with a cross, while other cardinal directions were denoted using letters or initials of wind names, i.e. the names of directions (Taylor 1951; Faričić et al. 2023a). The primary motivation for marking east with a cross was its association with Jerusalem, the direction where Jesus Christ was crucified. In a broader context, it reflected the tradition of orienting towards the east as a symbol of light (the rising sun)—*ex Oriente lux* (Gordon 1971; Healey 1977). In the Gospel of Luke, it is recorded that the elder Simeon, in the Jerusalem temple, referred to the infant Jesus as "a light for revelation to the Gentiles" (Luke, 2: 32). Jerusalem lies to the east of the locations where the earliest nautical charts were created. However, the use of the cross to signify east was not dependent on the specific placement of the compass rose on the chart, meaning that the cross was used regardless of whether Jerusalem was positioned directly east of the compass rose. On some nautical charts of the Adriatic, the compass roses featuring a cross to mark east spanned the entire field of the map. For instance, Joan Martines employed this approach on a nautical chart he created in Messina in 1550.

On some nautical charts, the east was marked with a cross on compass roses, which, based on their placement on the chart, pointed towards the Ottoman capital of Istanbul, one of the leading centres of Islam. Furthermore, the cross also indicated the direction of Istanbul and other Ottoman cities marked with a crescent moon, the symbol of the Ottoman Empire and Islam (see Chap. 5, Fig. 5.3). This raises the question of how such a contradictory situation could arise. Given the centuries-long practice of uncritical reproduction of older nautical charts, which led to a sort of "freezing" of their content, it is likely that at some unknown point, the direct association of the cross as a symbol of the east with the location of Christ's passion, crucifixion, death, and resurrection faded. Consequently, marking the east with a cross became an established practice in which the symbol assumed the function of a sign devoid of its primary meaning tied to the origins of Christianity. Instead, it acquired a secondary meaning as a mere indicator of the east, taken for granted by users. This evolution made it possible for a cross to point towards a crescent moon without either the cartographer or the user perceiving any confusion or contradiction in the symbolism.

While the cross on compass roses was consistently associated with the east, the crescent moon symbol was linked to both the south and the east in terms of cardinal directions. This symbol was closely tied to Islam and the Ottoman Empire, where Islam was the state religion. Considering that North Africa and the Middle East, predominantly Muslim regions, are situated south and east of European cartographic centres, marking the south and east with a crescent moon acknowledged this reality. However, this practice was rare on maps created by European authors. The substitution of the cross with the crescent moon as a marker for the east occurred much later, only on the cusp of modern nautical cartography. It appeared when the crescent moon was used in this role on compass roses in the *Carta di Cabottaggio del Mare Adriatico*, published by the Military Geographical Institute in Milan between 1822 and 1824 (Fig. 6.3).

3.2 Marking Territories with Religious Symbols

The use of the crescent moon and other religious symbols on nautical charts also served another purpose: distinguishing Adriatic regions based on their political affiliation. A prime example of such a nautical chart is the one created by Antonio Millo in Venice in 1583 (see Chap. 3, Fig. 3.2). Millo, a Venetian cartographer of Greek origin, was primarily a ship captain. As an experienced mariner, he understood the importance of recognizing the political landscape of the Adriatic, which, in the late sixteenth century, was a theatre of conflict between several Christian states and the Ottoman Empire. Millo marked the Adriatic regions under Ottoman control with flags featuring a crescent moon. Venetian territories were represented by a flag bearing the Winged Lion, a symbol of St Mark the Evangelist (referenced in the *Book of Revelation*, where one of the evangelists is symbolically depicted as a winged lion; Revelation 4:4–8). St Mark

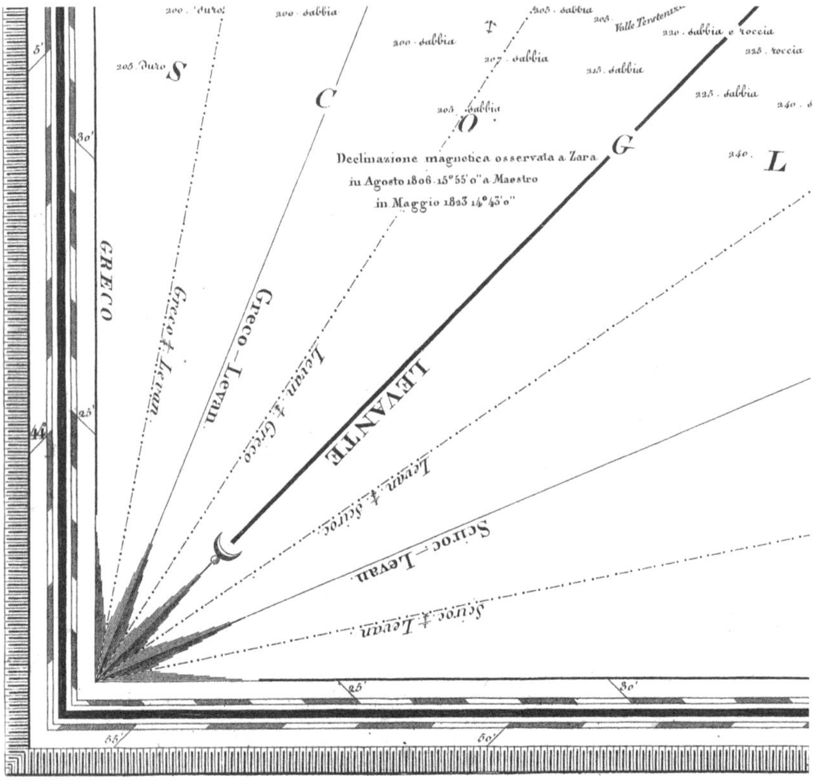

Fig. 6.3 Compass rose on which east is indicated by a crescent moon on the seventh sheet of the *Carta di cabotaggio del Mare Adriatico, 1822–1824* (Source: State Archives in Zadar, Cartographic collection, Zadar; HR-DAZD-383 No. 3.2)

was the patron saint of Venice (Frank 2019). The Republic of Ragusa (Dubrovnik) was marked with a flag depicting St Blaise, a bishop from Sebaste in Asia Minor, martyred by Roman authorities in the early fourth century and the patron saint of the small Adriatic state. The Papal States were represented by the Keys of Heaven (Keys of St Peter), symbolizing the papal authority as the successor of Apostle Peter and Christ's representative on Earth (Feuillet 2007). For other Adriatic states, Millo used heraldic elements unrelated to religion, such as the double-headed eagle symbolizing the Habsburg dynasty and two diagonally positioned towers representing the Kingdom of Sicily (also referred to as the Kingdom of Naples).

The heraldic symbolism of the two Adriatic republics—Venetian and Ragusan—underscores their deep identification with their respective patron saints, St Mark for Venice and St Blaise for Dubrovnik. These saints were not only viewed as protectors but became so integral to the republics' identities that the states were informally known as the "Republic of Saint Mark" and the "Republic of Saint Blaise". This devotion extended beyond symbolism; during periods of territorial expansion, new churches were constructed, or existing ones rededicated, in honour of these saints. Their images also prominently adorned secular buildings, reflecting their pervasive influence on civic and religious life (Wills 2002; Marinković 2009).

3.3 *The Winged Lion of St Mark: The Ubiquitous Symbol of Venice*

The Winged Lion of St Mark appeared on many nautical charts, often without directly referencing specific areas under the jurisdiction of the Republic of Venice. These charts, produced in Venice, reflected the cartographers' role as proponents of the Venetian political narrative, which portrayed Venice as the dominant power over the entire Adriatic. This narrative was conveyed through both visual and textual elements, including depictions of the Winged Lion, decorative allegorical illustrations of Venice as a sovereign ruler, and the use of the hydronym *Golfo di Venezia* to label the Adriatic Sea (Faričić et al. 2023b). For example, Alvise Gramolin incorporated the Winged Lion of St Mark into his 1624 nautical chart of the Adriatic and Ionian Seas, crafted in Venice (Fig. 6.4a). This narrative was similarly embraced by other European cartographers working in Venice. Notably, the French cartographer Jean François Rouissin featured the Winged Lion (Fig. 6.4b) prominently perched above a panoramic view of Venice in his 1661 map of the Adriatic Sea (*Carta du Golfo di Venetia*).

In addition to the previously mentioned examples where the figure of the Winged Lion of St Mark on nautical charts is associated with the Republic of Venice, one nautical chart features the lion alongside symbols of the other three evangelists. This chart of the central Mediterranean, part of an atlas of nautical charts by an unknown author, was created in the late fifteenth century (Fig. 6.5). On this map, St Mark is depicted as a winged lion, St Luke as a winged bull, St Matthew as a winged man, and St John as an eagle. Their names are inscribed next to these drawings in Latin. The depictions of the evangelists serve a decorative purpose and are

Fig. 6.4 Details featuring the depiction of the Winged Lion of St Mark: (**a**) Alvise Gramolin, Carta nautica del mar Adriatico, Venice, 1624 (Source: Museo Correr, Gabinetto di Cartografia, Cl. XLIVa, n. 0044); (**b**) Jean François Roussin, Carta du Golfo di Venetia, Venice, 1661 (Source: The Huntington Library, Art Museum, and Botanical Gardens, Library Collections, Maps and Atlases, Portolans, San Marino, CA, USA; mssHM 37)

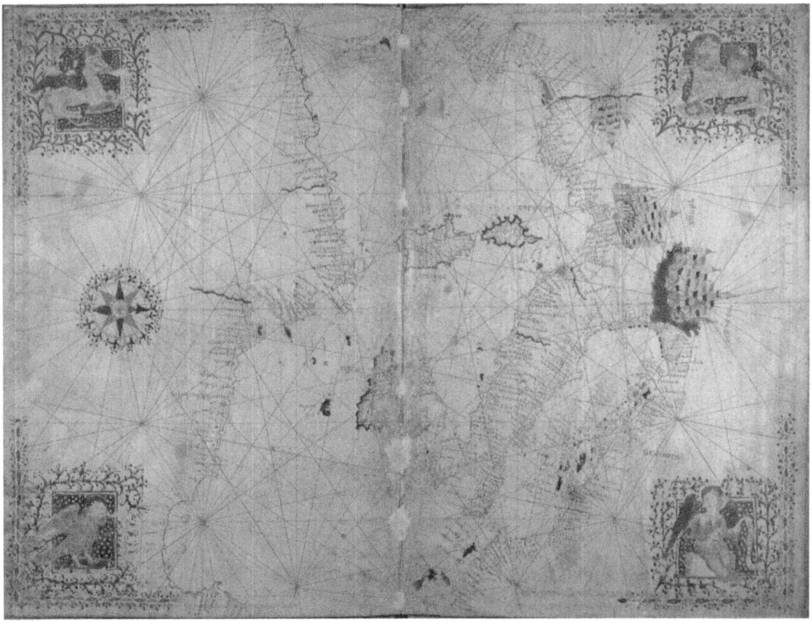

Fig. 6.5 Unknown Author, nautical chart of the Central Mediterranean, late fifteenth century (Source: *Biblioteca Nazionale Marciana*; It. IV, 9 (=5090), t. 2)

not linked to any specific political entity. Instead, they signify the Christian identity of the depicted area during a period when it was under threat from Ottoman expansion.

3.4 *A Synchronic Representation of Chronologically Diverse Religious Motifs*

The Winged Lion of St Mark was depicted on some maps in an unusual correlation with figures from ancient Greek and Roman religion and their associated mythology. These compositions do not reflect religious syncretism but align with Renaissance and Baroque traditions of referencing ancient scientific and cultural heritage. While this was not common on nautical charts, given the connection of motifs from ancient religion and mythology to the sea, geographical maps featuring religious symbols from the Christian and pre-Christian ancient world broadly belong to the maritime cartographic tradition. Examples of such maps depicting the Adriatic include the map of Italy by the German cartographer and engraver Mattheus Greuter, printed in Bologna in 1630, and the map of Dalmatia by the Venetian cartographer V. M. Coronelli, printed in Venice around 1690. On Greuter's map, Venice is portrayed allegorically as a sovereign female figure adorned with insignia of the Doge of Venice, accompanied by the Winged Lion, while facing them stands Poseidon, the Greek god of the sea (or Neptune in Roman mythology), recognizable by his trident (Fig. 6.6a). On Coronelli's map, Venice rides in a marine chariot alongside a priest holding a large cross and chalice, accompanied by angels, one of whom carries the flag of St Mark (Fig. 6.6b). Surrounding them are

a) b)

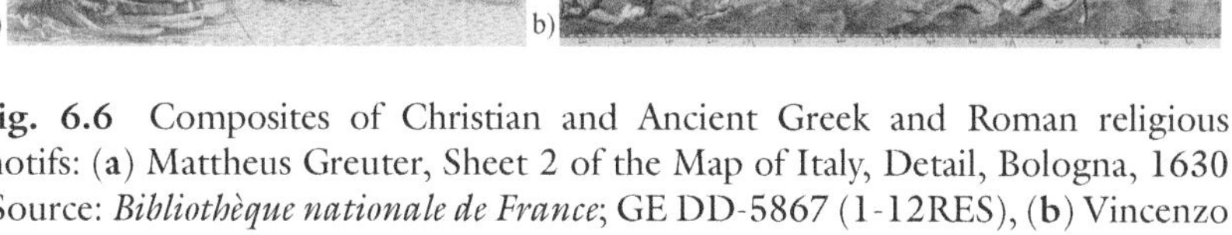

Fig. 6.6 Composites of Christian and Ancient Greek and Roman religious motifs: (**a**) Mattheus Greuter, Sheet 2 of the Map of Italy, Detail, Bologna, 1630 (Source: *Bibliothèque nationale de France*; GE DD-5867 (1-12RES), (**b**) Vincenzo Maria Coronelli, Map of the Dalmatia, Detail, Venice, ca 1690 (Source: Maritime and History Museum of the Croatian Littoral, Rijeka; PPMHP 122117)

mermaids and mermen, mythological children of the ancient Greek sea god Triton (the son of Poseidon and Amphitrite). These motifs, while secondarily referencing religion, primarily symbolize Venice's political and economic relationship with the sea. As a Catholic priest, Coronelli showed no hesitation in incorporating elements of ancient mythology into his artistic representations, despite their apparent contradiction with Church dogma. However, this dogma was not challenged, as the map depicting the Adriatic was not theological content subject to scrutiny by those responsible for safeguarding orthodoxy. Moreover, many church dignitaries of the time were collectors of ancient artefacts and patrons of visual arts, where figures and events from Greek and Roman religion and mythology were frequently depicted (Baker-Bates et al. 2019; Beaven 2019). Coronelli also expressed freedom of thought in other contexts. For instance, in his atlas *Corso Geografico Universale*, dedicated to Pope Clement XI in 1692, he included depictions of celestial movements that were still formally contested by Church authorities at the time, such as the motion of heavenly bodies in Copernicus's heliocentric system.

3.5 Angels on Nautical Charts

In addition to the religious symbols used in the heraldry of Adriatic states, other symbols also appeared on some nautical charts in a similar capacity. For example, on the 1642 nautical chart of the Adriatic Sea by Pierre Collin, created in Marseille, the Papal States were represented by the papal tiara, a symbol of papal authority and Rome as the centre of the Catholic Church (Fig. 6.7).

The same symbol was used on the 1720 nautical chart of the Adriatic Sea by Pieter Van der Aa, printed in Leiden (Fig. 6.8). However, the depiction of the papal tiara on Van der Aa's nautical chart serves an intriguing purpose. It is drawn above a cloud that hovers over a column topped by the Winged Lion, with angels at its base holding the coat of arms of Venice. The lion holds an open book inscribed with the text *Pax tibi Marce Evangelista Meus* ("Peace be with you, Mark, my evangelist"). To the left is the *Bucentaur*, the ceremonial galley of the Doge, from which the annual Wedding of Venice with the Sea (*Sposalizio del Mare*) was celebrated on the Feast of the Ascension (*Festa della Sensa*). This event, in addition to marking a religious feast, was also a grand state celebration

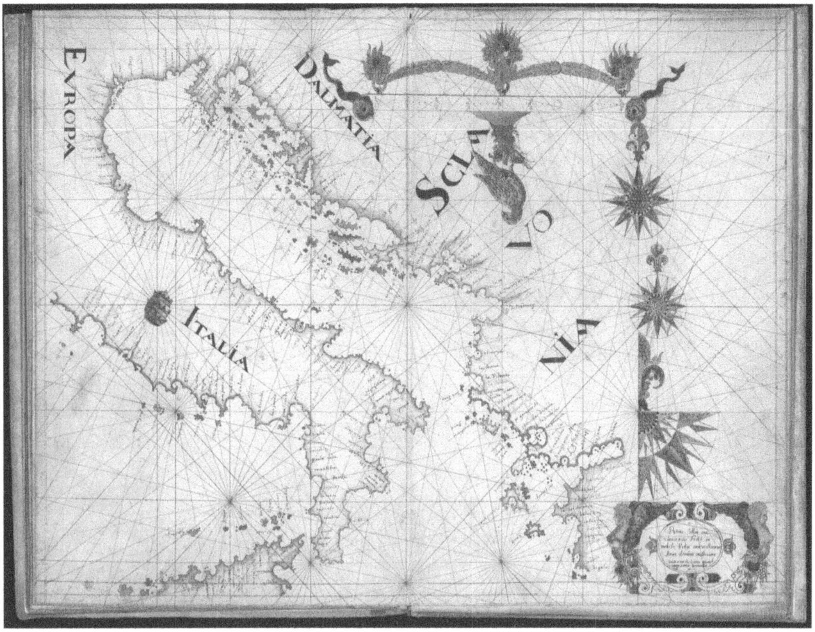

Fig. 6.7 Pierre Collin, nautical chart of the Adriatic Sea, Marseille, 1642 (Source: *Bibliothèque Municipale de Lyon*, Lyon; MS 177)

(Bertoncin 1997). According to Venetian tradition, Venice's dominion over the Adriatic was confirmed by Pope Alexander III in 1177 (De Vivo 2003). This state myth was not unsettled by the fact that many subsequent popes clashed with Venice over political disputes on the Apennine Peninsula (Ivetic 2019). By incorporating various religious motifs, including angels as messengers of God, the Dutch cartographer alluded to the political rights of Venice, suggesting that these rights were validated not only by the highest Church authority but also by divine authority. The triumphant atmosphere in Venice following the War of Morea is particularly evident in the depiction of a fallen flag bearing three crescents, symbolizing the defeated Ottoman Empire. Within the same symbolic context, one can also interpret the depiction of a falling turban—significantly, without the head it once adorned—emphasizing the disempowerment of the former adversary.

Fig. 6.8 Pierre Van der Aa, Detail from the chart Golfe de Venise avec les Côtes maritimes, Bayes et Ports etc. De la Gréce, Dalmatie et Italie, Leiden, 1720 (Source: *Croatian State Archives, Cartographic Collection*, Zagreb; HR-HDA-902, E.IV.29)

Angels, depicted as spiritual beings in the form of scantily dressed children with wings, were not solely "harnessed" to the cart of political aspirations on nautical charts of the Adriatic Sea. On a 1700 map of the Adriatic Sea printed in Amsterdam by Ioannes van Keulen, angels in the cartouche hold astronomical instruments (an astrolabe and Jacob's staff) and a celestial sphere map (Fig. 6.9). In this depiction, they symbolize astronomy, the science of the heavens, which in a spiritual sense represents God's domain. In Van Keulen's interpretation, the angels are a symbol of divine support for astronomical observations and cartography—scientific endeavours that hold great importance for navigation.

Fig. 6.9 Ioannes van Keulen, detail from the chart Pas-Caart van de Weder zytsche Zee-kusten soo van Italia als Dalmatia en Griecken Inde Golff van Venetien, Amsterdam, 1700 (Source: *Croatian State Archives, Cartographic Collection, Zagreb*; HR-HDA-902, E.IV.30)

4 Conclusion

The study of approximately one hundred nautical charts depicting the Adriatic region during the Middle Ages and early modern period uncovered a diverse and intricate array of religious symbols. These symbols were not used on all nautical charts, but on many they served as important decorative elements that also had a utilitarian function—either directly, in the form of compass roses and indications of the political affiliation of individual Adriatic regions, or indirectly, as expressions of belonging and power in a wider scope. Religious symbols on the old nautical chart of the Adriatic Sea can be categorized into two groups:

(a) primary or original symbols: these include decorative elements with explicit religious motifs, such as angels, Christian saints, and Greek mythological figures;
(b) secondary or intermediate symbols: these encompass symbols that indirectly carry religious connotations, such as the cross and crescent used to mark east on compass roses, and heraldic elements with religious motifs (e.g. the papal tiara representing the Papal States, and the Winged Lion of St Mark symbolizing the Republic of Venice).

Religious symbols on nautical charts were predominantly rooted in Christianity, but Islamic motifs occasionally appeared, reflecting the historical coexistence and conflicts among Mediterranean states associated with these two religions. The Adriatic Sea was a stage for complex interactions between the Ottoman Empire and the Adriatic states, primarily the Republic of Venice. The symbolic clash of the cross and crescent, as well as the Winged Lion of St Mark and the crescent, depicted on nautical charts, primarily represented the rivalry between imperial systems competing for political and economic dominance. Only secondarily did it reflect the turbulent interactions between followers of different religions.

The Renaissance revival of interest in ancient cultural and scientific paradigms, including Greek and Roman religions and mythologies, further enriched the iconography of these charts. This blending of influences led to synchronic representations of chronologically diverse content, resulting in unique composites of Christian and pre-Christian religious imagery. While often serving decorative purposes, nautical charts also conveyed complex and nuanced messages to sailors and other users, blending practicality with cultural and symbolic significance. This reflected the layered cultural sedimentation and ongoing exchanges in the Adriatic region, embodying its complex historical and cultural dynamics.

References

Astengo C (2007) The renaissance chart tradition in the Mediterranean. In: Woodward D (ed) The history of cartography, vol 3, Part 1 – Cartography in the European Renaissance. University of Chicago Press, Chicago, pp 174–262

Baker-Bates P, Hollingsworth M, Witte A (2019) Cardinals as patrons of the visual arts. In: Hollingsworth M, Pattenden M, Witte A (eds) A companion to the Early Modern Cardinal Brill, pp 511–534

Beaven L (2019) Elite patronage and collecting. In: Jones PM, Wisch B, Ditchfield S (eds) A companion to Early Modern Rome, 1492–1692. Brill, pp 387–411

Bertoncin M (1997) Dallo sposalizio del mare al baratto con la terra. Geostorie 5:6–13

Buttimer A (2015) Introduction. In: Buttimer A, Seamon D (eds) The human experience of space and place. Routledge, pp 13–18

Caldwell MC (2014) Flower of the lily: Late-Medieval Religious and Heraldic Symbolism in Paris, Bibliothèque Nationale de France, MS Français 146. Early Music Hist 33:1–60. https://doi.org/10.1017/S0261127913000119

Campbell T (1987) Portolan charts from the Late Thirteenth Century to 1500. In: Harley JB, Woodward D (eds) The history of cartography, vol 1 – Cartography in prehistoric, ancient and medieval Europe and the Mediterranean University of Chicago Press, Chicago, pp 371–463

Canney MA (1937) Stella maris. Revue de l'histoire des religions 115:90–94

Cartwright W (2009) Art and cartographic communication. In: Cartwright W, Gartner G, Lehn L (eds) Cartography and art. Springer-Verlag, pp 9–22

Chaline O, Doumerc B, Ducellier A, Sivignon M (2001) Histoire de l'Adriatique. Seuil

Cosgrove D (1992) Mapping new worlds: culture and cartography in sixteenth-century Venice. Imago mundi 44(1):65–89. https://doi.org/10.1080/03085699208592741

Cosgrove D (1999) Introduction: mapping meaning. In: Cosgrove D (ed) Mappings Reaktion Books, pp 1–23

De Vivo F (2003) Historical justifications of Venetian power in the Adriatic. J Hist Ideas 64:159–176. https://doi.org/10.1353/jhi.2003.0031

Edney MH (2005) The origins and development of JB Harley's cartographic theories. Cartographica 40(1-2):1–143. https://doi.org/10.3138/928H-787R-3087-5V02

Faričić J, Kljajić I, Mirošević L, Mlinarić D (2023a) Symbolism of compass roses on Early Modern Nautical charts of the Adriatic Sea. KN-J Cartography Geographic Inf73(1):19–37. https://doi.org/10.1007/s42489-022-00129-z

Faričić J, Selva O, Umek D (2023b) Geographical names of the Adriatic Sea on medieval and early-modern maps and nautical charts. J Hist Geography 82:68–80. https://doi.org/10.1016/j.jhg.2023.09.007

Farinelli F (1999) Storia e geografia dell'Adriatico. In: Turri E (ed) Adriatico Mare d'Europa: La geografia e la storia. Rolo Banca 1473, pp 16–24

Feuillet M (2007) Lessico dei simboli cristiani. Edizioni Arkeios, Rome

Forsberg T (2003) The ground without foundation: territory as a social construct. Geopolitics 8(2):7–24. https://doi.org/10.1080/714001038

Frank M (2019) Representing the Republic in Seventeenth-Century Venice. Radovi Instituta za povijest umjetnosti 43:113–122. https://doi.org/10.31664/ripu.2019.43.09

Gambin T (2014) Maritime activity and the divine - an overview of religious expression by Mediterranean seafarers, fishermen and travellers. In: Agius DA, Gambin T, Trakadas A (eds) Ships, saints and sealore: cultural heritage and ethnography of the Mediterranean and the Red Sea. Archeopress Archeology, pp 3–12

Gaspar JA, Leitão H (2018) What is a nautical chart, really? Uncovering the geometry of early modern nautical charts. J Cult Heritage 29:130–136. https://doi.org/10.1016/j.culher.2017.09.008

Gaspar JA, Leitão H (2019) Early Modern Nautical charts and maps: working through different cartographic paradigms. J Early Modern Hist 23(1):1–28. https://doi.org/10.1163/15700658-12342627

Goodrich TD (2004) A Cartographic Innovation of Piri Reis in His Kitab-i Bahriye. In: Baykara T (ed) CIÉPO XIV. Sempoziumu Bildirileri, Türk Tarih Kurumu Basımevi, pp 201–210

Gordon BL (1971) Sacred directions, orientation, and the top of the map. Hist Religions 1(3):211–227. https://doi.org/10.1086/462630

Harley JB (1989) Deconstructing the map. Cartographica 26:1–20. https://doi.org/10.3138/E635-7827-1757-9T53

Healey T (1977) The symbolism of the cross in sacred and secular art. Leonardo 10(4):289–294. https://doi.org/10.2307/1573764

Holloway J, Valins O (2002) Editorial: placing religion and spirituality in geography. Social Cult Geography 3(1):5–9. https://doi.org/10.1080/14649360120114107

Ivetic E (2019) Storia dell'Adriatico: un mare e la sua civiltà. Società editrice il Mulino

Jackson RH, Henrie R (1983) Perception of sacred space. J Cult Geography 3(2):94–107. https://doi.org/10.1080/08873638309478598

Karssen AJ (1980) The artistic elements in map design. Cartographic J 17(2):121–127. https://doi.org/10.1179/caj.1980.17.2.124

Katić M, Vedriš T (2023) The historical context of boat processions in Adriatic Maritime Pilgrimages. Religions 14(7):884. https://doi.org/10.3390/rel14070884

Katičić R (1968) Najstariji jezici i narodi u senjskom primorju. Senjski zbornik: prilozi za geografiju, etnologiju, gospodarstvo, povijest i kulturu 3(1):46–53
Kisić A (1999) Maritime relations between the Republic of Dubrovnik and Malta: Votive paintings of Ragusan Seamen in Malta. J Mediterranean Stud 9(1):64–73. https://muse.jhu.edu/article/670243
Kitchin R, Dodge M (2007) Rethinking maps. Progr Hum Geography 31(3):331–344. https://doi.org/10.1177/0309132507077082
Kozličić M (1990) Historijska geografija istočnog Jadrana u starom vijeku. Književni krug
Krygier JB (1995) Cartography as an art and a science? Cartographic J 32(1):3–10. https://doi.org/10.1179/000870495787073762
Marinković A (2009) Territorial expansion of the Ragusan Commune/Republic and the Churches of its patron saints. Dubrovnik Annals 13:7–23
Maqbul Ahmad S (1992) Cartography of al-Sharīf al-Idrīsī. In: Harley JB, Woodward D (eds) The history of cartography, vol 2, Part 1, Cartography in the traditional Islamic and South Asian Societies. The University of Chicago Press, pp 156–174
Meyer D (2008) Apollonius as a Hellenistic geographer. In: Papanghelis D, Rengakos A (eds) T. Brill's companion to Apollonius Rhodius, Brill, pp 267–285
Milanesi M (2016) Vincenzo Coronelli Cosmographer (1650-1718). Brepols
Paasi A (2000) Territorial identities as social constructs. Hagar – Int Soc Sci Rev 1(2):91–113
Park C (1994) Sacred worlds – an introduction to geography and religion. Routledge
Ribeiro DM, Caquard S (2018) Cartography and art. In: Wilson JP (ed) The geographic information science & technology body of knowledge. UCGIS, Ithaca
Santoro NJ (2011) Mary in our life: Atlas of the names and titles of Mary, the Mother of Jesus, and their place in Marian devotion. iUniverse
Schilder G, Egmond M (2007) Maritime Cartography in the Low Countries during the Renaissance. In Woodward D (ed) The history of cartography, vol 3, Part 2 – Cartography in the European Renaissance. The University of Chicago Press, pp 1384–1432
Sheehan KE (2012) Utility and aesthetic: the function and subjectivity of two fifteenth-century portolan charts. Portolan 83:7–23
Sopher DE (1967) Geography of religions. Prentice-Hall
Soucek S (1992) Islamic Charting in the Mediterranean. In: Harley JB, Woodward D (eds) The history of cartography, vol 2, Part 1, Cartography in the traditional Islamic and South Asian Societies. The University of Chicago Press, pp 263–292
Stump RW (2008) The geography of religion – faith, space and place. Rowman and Littlefield Publishers

Summa M (2017) Phenomenology of imaginal space. In: Catena MT, Masi F (eds) The changing faces of space. Springer Nature, pp 75–99
Taylor EGR (1951) Early charts and the origin of the compass rose. J Navigation 4(4):351–356. https://doi.org/10.1017/S0373463300034214
Timmermann A (2021) The ship in the shop: an art history of late medieval ship models. Int J Maritime Hist 33(2):257–288. https://doi.org/10.1177/08438714211013534
Vagnon E (2021) Representations. Painting, mapping, re-inventing the seas of the world. In: Lambourn E (ed) A cultural history of the Sea II. Bloomsbury Publishing, pp 159–192
Wills G (2002) Venice: Lion City: The Religion of Empire. Simon and Schuster, Washington

CHAPTER 7

Credibility of the Maps of the Adriatic Isolarii in the Sixteenth Century: Creation of Different Geographical Identities

Milorad Pavić

1 Introduction

Isolarii, or literally "books of islands" or "depictions of islands," represent books with descriptions of islands (Tolias 2007; Pavić 2013). The earliest such work was *Insullae Archipelagi*, written in Florence in 1420 by the Franciscan priest and traveller Cristoforo Buondelmonti. Although the genre was originally established in Florence, at that time the main cultural centre of the European Mediterranean and the Italian area, from the second half of the fifteenth century, Venice became the main centre of isolarii's production, a rule it maintained for three centuries.

If *portolans* or *portolani*—the late medieval and early modern inscriptions containing bearings and distances between ports, as well as depths and anchorages—could be described as navigation guides similar to modern rutters (Marelić 2022), the isolarii could be likewise described as travel guides. The cartographic representations of the islands and parts of the

M. Pavić (✉)
Department of History, University of Zadar, Zadar, Croatia
e-mail: mpavic@unizd.hr

J. Faričić, T. Marelić (eds.), *Early Modern Nautical Charts of the Adriatic Sea*, https://doi.org/10.1007/978-981-96-8046-7_7

coastland, made for the first time to a considerably large map scale, combining nautical and geographical elements, were accompanied by textual content referring to economic and political phenomena, as well as excerpts referring to historical events. In the later stages of isolarii production, the coastal area progressively occupied a greater portion of the map framework, resembling panoramic views and regional maps. The demand for such works grew as a result of the significant reader interest at the time, and, like the rest of the cartographic production, their production was converted to printing techniques in the second half of the fifteenth century, with the earliest printing tables created using woodcut techniques. Such a process required the engraver to carve the entire map content inverted (mirrored) into a woodblock, whereas the segments that were not printed clearly enough or had been too detailed were added or corrected manually afterward (Bliss 2013). *Libro di Benedetto Bordone: nel qual si ragiona de tutte l'isole del mondo, con li lor nomi antichi & moderni, historie, fauole, & modi del loro uiuere, & in qual parte del mare stanno, & in qual parallelo & clima giacciono*[1] made in 1528 represents the most renowned such work made using the woodcarving technique. Since the late sixteenth century, the copperplate engraving printing technique began to be used more extensively, giving a new impetus to the development of the genre. The technique allowed the mapmaker to trace the map content with engraving tools on a polished 2–5 mm thick copperplate. In addition, the carved panel was coated with paint, after which the paint was carefully removed from the surface. By doing so, the paint that remained within the plate's grooves could have been transferred (printed) on paper after it was put into a press (Hind 1923).

As a result of the augmentation of geographical and cartographic knowledge, further progress was made in detailing the representations of coastal features. Some former standards of cartographic generalisation, such as the stylised arcs representing the coastlines on portolan charts (Campbell 1987), were abandoned, and additional efforts were made to depict the coastline and mainland landscapes in more detail. The maps contained in isolarii contained a large number of toponyms and presented other elements such as depictions of battles or the current political situation. The spirit of the Renaissance was ubiquitous; the maps were

[1] English translation: *Book of Benedetto Bordone: in which we reason of all the islands of the world, with their ancient & modern names, histories, fables, and ways of life, and in which part of the sea they are, and in what parallel and climate they lie.*

becoming more refined, detailed, and decorative. The accompanying textual elements, which were not necessarily integral parts of the isolarii—sometimes written on cartouches and sometimes made as separate texts—were added to complement their cartographic content with additional information. The most renowned authors of isolarii from that period were Tomaso Porcachi, Giovani Francesco Camocio, Simone Pinargenti, Giacomo Franco, Alonso de Santa Cruz, Antonio Millo, Pîrî Reis, Giuseppe Rosaccio, and André Thevet (Novak et al. 2005; Faričić and Simičić 2005; Pavić 2013; Kljajić and Razum 2016; Faričić et al. 2020; Faričić and Lončar 2024).

Although there was a certain stagnation in the development of the isolarii during nearly the entire seventeenth century, a significant flourishing occurred at its end through the works of the Venetian geographer and cartographer Vincenzo Maria Coronelli. His fruitful geographical works earned him the title of chief geographer of the Republic of Venice in 1685, enhancing his scholarly reputation and granting him access to all available information from the state archives. Coronelli travelled extensively, collaborated with the established geographers and publishers of his time, and worked on the sporadic local geographical survey of the coastal areas. For the territorial acquisitions that were not controlled by the Republic of Venice, he was supported by the Austrian authorities. Coronelli's numerous works reflect nearly the entire geographical knowledge of the Venetians regarding the Adriatic Sea and the surrounding areas. His efforts to overcome certain shortcomings of the isolarii and to consolidate the existing geographical knowledge resulted in his work *Mari, Golfi, Isole, Spiaggie, Porti, Citta, Fortezze, Ed altri Luoghi Dell'Istria, Quarner, Dalmazia Albania Epiro, e Livadia, Delineati, e Descritti Dal P. Generale Coroneli*,[2] the so-called Small isolario (Coronelli 1694) published in 1694 in Venice. The highly detailed fortification plan he provided, based on his surveying work, attracted special attention. In the absence of more modern data, he often used the results of his predecessors, so his descriptions and text are not completely devoid of inaccuracies and superficiality. In the following decades, the interest in isolarii further declined, and the genre was never revived.

[2] English translation: *Seas, Gulfs, Islands, Beaches, Ports, Cities, Fortresses, and other Places of Istria, Quarner, Dalmatia, Albania, Epirus, and Livadia, Delineated, and Described by Father General Coroneli.*

Throughout the entirety of their production, isolarii were characterised by certain limitations and aggravations—such as the insufficient geographical knowledge about the depicted areas, strong reliance on sources from classical antiquity, inadequate complementarities between cartographic and textual traditions, technological infeasibility of conducting field surveys, and excessive market demands—that considerably reduced the credibility of their geographical content, which their authors were unable to resolve. This will be discussed in more detail in the following paragraphs.

2 Inaccuracy of Maps as a Result of Distorted Map Content

The inaccuracy of maps contained in isolarii was analysed from six different approaches. The first three relate to how and why the islands and parts of the mainland were drawn due to stylised renderings of coastlines, disproportionate depictions of coastline features, and distorted dimensions of the mapped islands. The fourth approach focuses on the pragmatic aspects of mapmaking, whereas the last two discuss the inconsistencies in multiple mappings of the same area by the same authors and some drawbacks that resulted from strong reliance on earlier-made maps that were typically used as templates due to a lack of genuinely observed spatial data.

2.1 Stylised Renderings of the Coastline

During the fifteenth and early sixteenth centuries, the small amount of available geographical knowledge about the island and coastal areas of the eastern Adriatic put cartographers in a somewhat unfavourable position. Their knowledge about the visual appearance and dimensions of the island and mainland coastline was primarily derived from their renderings on portolan charts, which, at the time, had an almost two-century tradition (Campbell 1987; Astengo 2007). Cartographers also had access to portolani texts, which seemingly supplemented their knowledge of the marine areas. Although the copper engraving technique for printing was already established, the authors of isolarii employed woodcut techniques until the end of the sixteenth century, presumably due to their relatively limited understanding of coastal features, which led to significantly generalised cartographic representations that did not necessitate more intricate engraving techniques. During that period, the coastlines in isolarii were depicted in a conventional portolan chart style, primarily represented as continuous arcs. Due to the woodcut technique, these arcs were engraved as relatively thick lines, which exaggerated their appearance and created the illusion that the geographical features were larger than they actually are. One of the most renowned authors of the woodcut-made isolarii is Benedetto Bordone, who made six maps of the Adriatic islands in his *Isolario Nel qual si ragiona di tutte l'Isole del mondo...*, produced in 1547. His map, *Parte de Schiavonia* (Fig. 7.1), for example, shows the island and mainland parts of the Kvarner Gulf and North Dalmatia with the islands of Krk (*vegia*), Cres (*cherso*), Rab (*arbe*), and Pag (*pago*), whose coastal contours strongly differ from their actual forms. Despite those shortcomings, Bordone's approach influenced many subsequent cartographers.

2.2 *Disproportionate Depictions of Coastline Features*

During its historical development, cartography often faced two main challenges. One challenge was how to successfully represent the spatial content on the flat surface of a map. The second one was the suitable extent of cartographic generalisation, which involves the selection of cartographic

Fig. 7.1 *Parte de Schiavonia* map from Benedetto Bordone's *Isolario* (1547) showing the island and mainland parts of the Kvarner Gulf and North Dalmatia (Source: University of Zadar, Research Library, R 317)

content to meet the specific needs of the map user. William Cartwright (2010), for example, categorises these demands by the terms "whereness", "whatness", and "whyness", the third being used to explain the motives behind the cartographer's decisions.

In discussing early modern cartography, it is essential to recognise the existence of two distinct traditions of mapmaking. One was more academic and grounded in the mathematical approach that arose from the rediscovery of Claudius Ptolemy's *Geography* in Western Europe during the early Renaissance. The other originated from the portolan chart tradition of the late medieval period and was more pragmatic (Harley 2001), initially adopted by the authors of the isolarii. In instances of inadequate geographical data at their disposal, they frequently consulted the works of other authors. When cartographers illustrated regions with which they were not adequately familiar, they frequently prioritised local social or political phenomena, simultaneously neglecting the topography of the represented areas. John Brian Harley (2011) introduces the concept of "deconstruction of the map" to highlight not only what is depicted on the map but also what is excluded. The appropriate example that illustrates such a phenomenon is Giuseppe Rosaccio's map of the island of Pag, published in his isolario *Viaggio da Venetia a Costantinopoli*..., published in 1598 (Fig. 7.2). The map does not contain any information related to its orientation, and its cartographic representations of its coastlines, illustrated as indented and rocky, are quite generalised. Moreover, the map merely depicts the island as an oblong semi-ring, with the Pag town at its centre, while in reality, it boasts one of the most complex coastline shapes on the eastern Adriatic, featuring numerous elongated bays and capes. In a manner akin to Rosaccio, other isolarii authors, including Antonio Millo (1582) and André Thevet (1586), portrayed the island of Pag as primarily compact in form and entirely devoid of certain northwestern geographical features. This suggests that these cartographers may have had knowledge solely of its southwestern region or that this area was their exclusive focus (Faričić et al. 2020). The town of Pag was encircled by walls featuring nine towers and, in accordance with the customs of contemporary Venetian towns, was fortified more robustly on the landward side than on the seaward side. Additionally, the seaside featured five gates, with only one located on the land side, as depicted in the veduta of Pag, through which access to the town was granted from the island after traversing a wooden bridge. Located south of the town in the bay are the Pag salt pans (*Saline*). Salt constituted the island's paramount wealth and principal source of

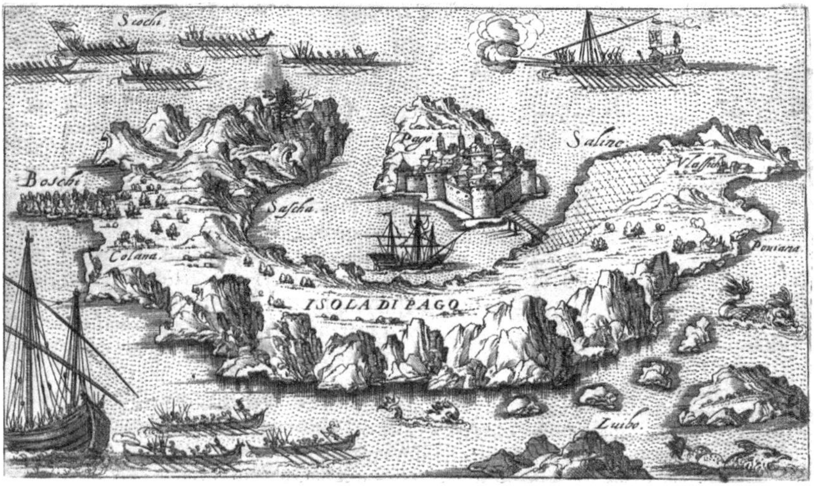

Fig. 7.2 The map *ISOLA DI PAGO* published in Giuseppe Rosaccio's isolario *Viaggio da Venetia a Costantinopoli Per Mare, e per Terra, insieme quello di Terra Santa*, Venice, 1598 (Source: National and University Library, Zagreb, Collection of Manuscripts and Old Books, Zagreb; R-VI-16°-32)

income. Despite the insufficient information regarding the town's actual appearance and layout, the depiction remains significant as it is the sole known representation of the sixteenth-century settlement of Pag. In the northern region of the Bay of Pag, the toponym *Sascha* is recorded, signifying the village of Caska, and it also encompasses the island of Pag's oldest designation (*Cissa*). On the western side of the island, the toponym *Colana* denotes the village of Kolan, while the southern region features the settlements of Vlašići (*Vlassich*) and Povljana (*Pouiana*). The island's interior is illustrated as largely barren, with the exception of a slender strip of land stretching to the northwest, which is, according to Rosaccio, covered in forests (*Boschi*). Rosaccio's methodology is warranted by his endeavour to represent the Bay of Pag and the town with utmost fidelity. Adjacent to the coast of Pag, opposite the island of Olib (*Luibo*), a cluster of islets is depicted (presumably representing the islands of Škrda, Maun, and Brušnjak), with the largest, located along the right margin of the map, most likely resembling the island of Vir. The representation of mythological marine creatures aims to infuse vitality into this section of the illustration, despite the overall cartographic work not appearing monotonous in any respect. The naval conflict between the Venetians and the Uskoks (*Scochi*) is illustrated in art, akin to the earlier map of the island of Rab. The upper section of the map depicts Uskok ships (ormanice) alongside a Venetian galley, whereas the lower left corner illustrates a *trabakul* sailboat (traditional type of cargo vessel) under assault by the Uskoks. This representation accurately reflects the military-political circumstances in the Velebit Channel during that period and unequivocally demonstrates the resolve of the opposing factions in their confrontation.

2.3 Distorted Depictions of Island Dimensions

The issue of the distorted perception of the Adriatic islands during the early modern period was thoroughly examined by Milorad Pavić (2012); thus, an attempt will be made to encapsulate the principal tenets of this issue. This matter presents a conflict between two concepts, each attempting to articulate the problem from distinct imagological and epistemological perspectives. Examining the issue in this manner, two essential enquiries emerge: First, how were erroneous perceptions regarding the island's dimensions formed, and second, to what degree were these perceptions validated and regarded as plausible within the context of the prevailing geographical understanding of the Adriatic region?

The dimensions of the island were typically conveyed through its length, width, and the extent of its perimeter. The length of an island refers to its longest dimension, typically measured between the two furthest points when the coastline exhibits minimal directional variation, while its width is defined as the distance approximately perpendicular to this length. The perimeter's extent, usually determined by the duration required to fully circumnavigate the island, was dictated by the ship's navigation at a minimum safe distance, hindering sailors from closely observing and documenting the island's coastal contours in great detail.

In the early modern period, Hvar was regarded as the largest island in the eastern Adriatic for a prolonged period, despite being the fourth largest, following Cres, Krk, and Brač; this perception likely stemmed from its length-to-width ratio exceeding 6:1. Consequently, the island of Cres was regarded as the second largest, and, in the eighteenth century, it 'emerged' as the largest Adriatic island. Conversely, the island of Brač, owing to its slightly oval configuration, has historically been perceived as significantly smaller than it actually is, while the island of Krk has been similarly underestimated due to its somewhat polygonal form. However, the irregularities in the shape of islands were not the sole factors influencing the perception of their size at that time. These perceptions were also shaped by classical antiquity sources, particularly during the Renaissance, and by the established maritime routes that favoured the island of Hvar over Brač (Kozličić 1990; Pavić 2003). Figure 7.3 illustrates the schematic example of how more extreme ratios between the perceived length and width of an island may have influenced the perception of its size. Similarly, the same effect is attained if a circle and an ellipse are utilised in place of quadrilaterals or any other more compact and elongated shape of equivalent area.

It is essential to note that during the initial years of isolarii production, the skewed perception of the Adriatic islands' dimensions was primarily influenced by a pragmatic approach, bolstered by the sailing distances documented in portolani and classical antiquity sources. At that time, the length of an island was not regarded as the shortest distance between its most distant points, referred to as the as-the-crow-flies distance, but was determined by the time required to navigate from one point to another along a curved route, whether by boat or overland via paved roads (Büsching 1778; Formaleoni 1786). This perception evolved over time due to the implementation of systematic and extensive geodetic and hydrographic surveys, alongside advancements in instruments used to observe the spatial data in the field. This culminated in the production of planimetrically accurate topographical maps and nautical charts, which depicted the island with high accuracy, allowing users to objectively ascertain their dimensions for the first time.

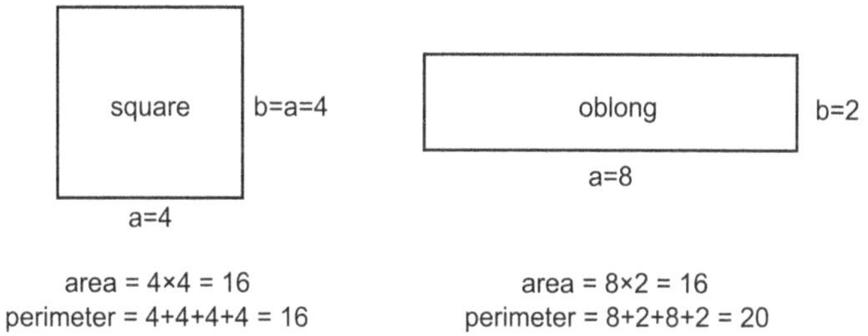

Fig. 7.3 The distinction between the two quadrilateral polygons possessing identical areas, yet differing in the length-to-width ratio of their sides (a:b)—the polygon on the right exhibits a longer perimeter, potentially giving the impression of being larger when measured by the time required to circumnavigate it (if it were an island)

2.4 *Pragmatism-Induced Inaccurate Representations of Islands*

Alongside the aforementioned distorted perceptions of the sizes of islands, some pragmatical reasons also played a role. The deconstruction of the map of the Zadar Region contained in Giovanni Francesco Camocio's isolario (1571), originally made by Martin Rota Kolunić (Fig. 7.4), reveals its significant importance not only from the strict cartographic point of view but also from a social, political, and maritime perspective. As has already been discussed in more detail in Pavić (2012), the depiction of a single, highly elongated island labelled *Isola di Zarra* that stretches approximately between the cities of Zadar (*Zarra Citta*) and Biograd (*Zara uechia*), separated from them by a channel called Zadar Channel

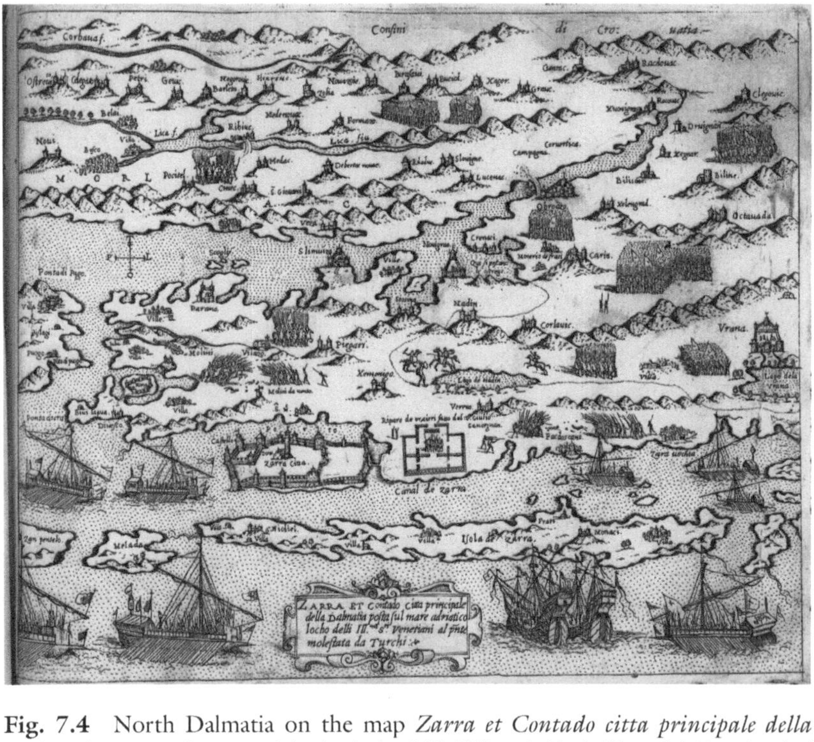

Fig. 7.4 North Dalmatia on the map *Zarra et Contado citta principale della Dalmatia posta sul mare adriatico locho delli Ill[ustrissi]mi S[igno]ri Venetiani al p[rese]nte molestada da Turchi* from Giovani Francesco Camocio's isolario, Venice, 1571 (Source: Archaeological Museum in Split; 45-e-39)

(*Canal de Zarra*), is reasonable from the navigational standpoint. In reality, that area is occupied by two islands called Ugljan and Pašman, divided from each other by a narrow and shallow passage (Ždrelac) that was, at the time, unsafe for navigation, and it is reasonable to assume that the cartographer decided to depict both islands as a single landmass in order to prevent sailors from entering such a potentially hazardous route. In comparison, the nautical charts of the same region made in the portolan chart fashion to a larger scale by the Ottoman navigator, geographer, and cartographer Pîrî Reis for his work *Kitab-ı Bahriye* (*Book of the Sea*) clearly show that the fact that Ugljan and Pašman were two separate islands was well known at the time, even to cartographers coming from more remote regions (Novak et al. 2005; Faričić and Simičić 2005).

A similar conclusion can be made regarding Camocio's image of the island of Vir (*Punta dura*), which is depicted as a cape, whereas in reality it is an island in close proximity to the mainland, divided only by a narrow passage called Privlački gaz. Etymologically, its Italian name, *Puntadura*, literally meaning a sharp tip, reflects the long-term perception of it being a cape, not an island, which is reasonable because Privlački gaz is only 450 meters wide and only around two meters deep. Such reasoning is supported by the testimonies of certain renowned eighteenth-century travelogue writers who claimed that the sea from the coast to the island can be crossed by foot (Fortis 1774). However, its navigational purpose is considerably challenged by the erroneous orientation of the crosshair drawn by the author below the Velebit Mountain (*MORLACA*). Namely, the axis of the Adriatic Sea is predominantly NW–SE, especially in the Zadar region, meaning that the image on Camocio's map is, in fact, rotated 45° anticlockwise, not 90° anticlockwise, as the crosshair indicating the cardinal directions suggests. In other words, the image of the crosshair itself should not have been aligned with the map's margins but tilted 45° anticlockwise.

2.5 *Inconsistencies in Multiple Mappings of the Same Area by the Same Authors*

Many scholars, especially historians and geographers, often rely on certain cartographic features illustrated on old maps or nautical charts. From a historical point of view, the content of every old map is conditioned by some characteristics of the period of its creation, and some of their elements were made with purely decorative purposes in mind. Therefore, they should not be approached superficially but put within the proper historical framework as objectively and accurately as possible, depending on the disposable number of other historical sources. This is especially important in cases of comparative analyses of two or more old maps of the same area made by the same author. The example in Fig. 7.5 shows how the same geographical feature—the church of St George on Cape Marjan in the Split area in this case—was depicted differently on two maps of Trogir and Split and areas, both of which are bound in Giuseppe Rosaccio's *Viaggio da Venetia, a Costantinopoli Per Mare* ... isolario (1598 and 1606). On the map of Trogir and its surrounding area (Fig. 7.5a), it is drawn as a three-nave church, complementary to the church of St James on Cape Rota on the island of Čiovo (Babić 1984), whereas on the

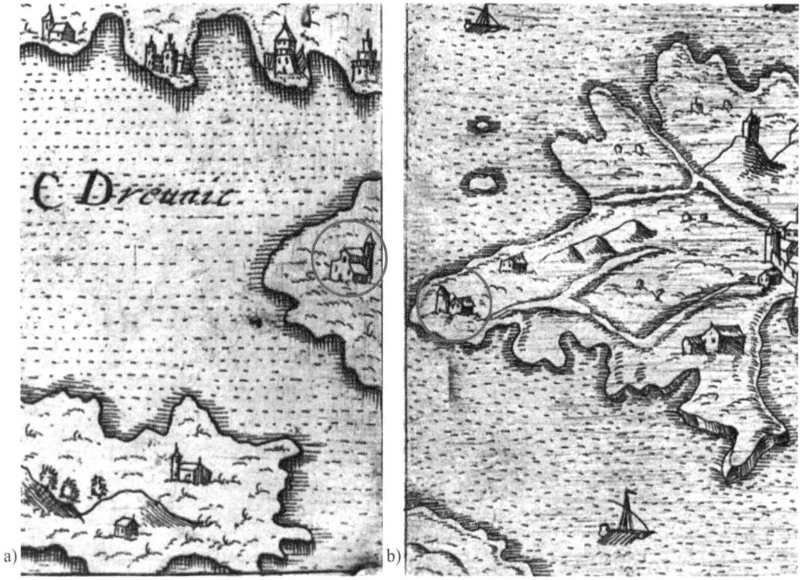

Fig. 7.5 The comparison of the Church of St George on Cape Marjan (encircled in red) on two different maps from Giuseppe Rosaccio's *Viaggio da Venetia, a Costantinopoli Per Mare ... isolario* (1606): (**a**) segment of the map of Trogir and its surroundings on the left; and (**b**) segment of the city of Split with its surroundings on the right (Source: National and University Library, Collection of Manuscripts and Old Books, Zagreb; R-VI-16°-32)

segment of the map of the Split micro-region (Fig. 7.5b), the same church is shown as a single-nave church.

The church of St George is, in fact, a single-nave stone building with a semicircular apse to the east (Prijatelj 1951), which confirms that the depiction of the church on the map of Trogir has no historical foundation. It is significant that among the geographical objects on the maps, churches are the most predominant manmade structures, whether in the centres of towns, on isolated capes, or at crossroads. They are not only built by peasants as their parish churches but also by city governments, church institutions, and wealthy individuals, representing the spatial focal points important for local identity and for spatial orientation. Therefore, the churches and their illustrations on nautical charts and maps were not only limited to a social dimension, but they were also important landmarks for coastal navigation in the Adriatic Sea.

2.6 Certain Drawbacks of Relying on the Earlier-Made Templates

Authors of nautical charts and maps in the isolarii frequently utilised the works of other authors as templates for their own cartographic creations, a common practice of the era that was not deemed unethical regarding the plagiarism of intellectual property. Also, a vast majority of those cartographers never had a chance to observe the mapped areas in person, so it is unreasonable to expect that they could have possessed the authentic geographical understanding of all the areas mapped in isolarii, shown in the example of Rosaccio's map of the island of Korčula (*CRVZOLA*) (Fig. 7.6). There is no direction of north on the map, meaning that at a superficial level one could assume that the map has been made in a typical "north-up" orientation. However, the map's approximate orientation is "south-up" for three reasons. Initially, he charted the town of Korčula in the "lower" section of the island, despite its actual position being in the north-eastern region. Secondly, the Korčulanski kanal (*Canal de Curzola*), which extends between the islands of Korčula and Hvar (to the north of Korčula), is inaccurately depicted as being situated "below" the island on his map. Thirdly, the extensive landmass that has been partially charted "beneath" the island, designated as *Augusta*, pertains to an incorrect geographical feature. The name refers to the island of Lastovo (commonly referred to as *Lagosta* on portolan charts), which is significantly smaller than Korčula and situated approximately 13 kilometres to its south. The illustrated

Fig. 7.6 The island of Korčula in Giuseppe Rosaccio's isolario, 1606 (Source: National and University Library, Collection of Manuscripts and Old Books, Zagreb; R-VI-16°-32)

landmass pertains to the Pelješac Peninsula (often labelled *Sabioncello* on old maps and charts), which is comparable in size to Korčula and situated less than two kilometres to the northwest.

The similarity between Rosaccio's map and a map of Korčula published in Camocio's isolario in 1571 (Fig. 7.7) is so striking that one gets the impression that he used Camocio's as a template. For example, coastline renderings on both maps are almost identical; the islets on the "upper" side of Korčula are also shown in the same way, as well as some other details, including the cartographic signs of two shipyards located near the town of Korčula—one "to the left", and the other "to the right" of it. In contrast to Rosaaccio's rather vague representation of the island, Camocio's map is more detailed; it is explicitly orientated "south up, north down", and the partial landmass drawn in its bottom part carries the correct toponym *Sabioncello* (the Pelješac peninsula). Camocio's map of the Koručula island was seemingly used as a template by French cartographer André Thevet in his *Le Grand Insulaire et Pilotage* isolario made in 1586 (Faričić et al. 2020).

In Rosaccio's *Viaggio*..., there are several maps of the eastern Adriatic coast that contain incorrectly inscribed toponyms. Also, on his map of the

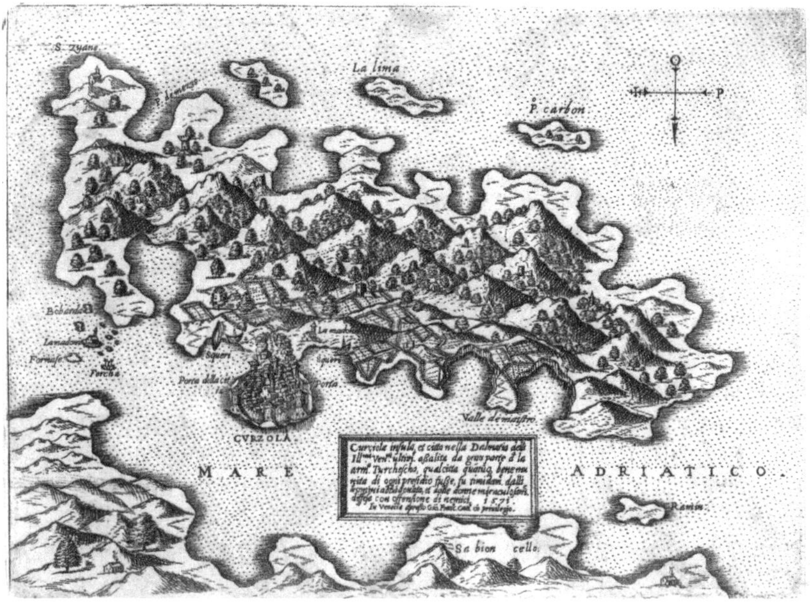

Fig. 7.7 Giovanni Francesco Camocio's map of the island of Korčula (1571), most likely used by Giuseppe Rosaccio as a template (Source: National and University Library, Collection of Maps and Atlases, Zagreb; SJZXVI-40)

Istrian peninsula, all the toponyms on its western coast are shifted and therefore incorrect because he failed to annotate two localities. The author of that map, most likely Andrea Bertelli, according to the autograph in the lower right corner of the map (Lago and Rossit 1981), most likely possessed some more accurate earlier-made maps and nautical charts of the Istrian peninsula, so perhaps there would have been an opportunity to avoid such mistakes if a more careful approach had been pursued. Also, on his map that depicts the city of Trogir and its surroundings (Fig. 7.5a), the toponym denoting the Drvenik Channel is incorrectly placed east of Trogir instead of west, and it was inscribed incorrectly (*C Dreunic* instead of *C Druenic*). Most likely, these errors occurred unintentionally, as the artist carved all the map content in a mirrored manner, which typically considers the plate's inverted east-west orientation. This is a skilful task that requires a high level of concentration and precision. Otherwise, it is easy to misjudge which elements (letters included) should be carved more to the left or to the right.

3 Discrepancies Between Toponyms on the Maps and in the Texts

Key elements in the depiction of geographical features in the isolarii, both in their maps and accompanying texts, are toponyms, which distinctly reflect the human organisation of the surrounding environment. Nonetheless, the toponyms on the maps of the isolarii do not consistently correspond with those in their accompanying texts, as shown in Table 7.1, containing a list of toponyms in isolarii created by Benedetto Bordone (1547), Giuseppe Rosaccio (1598), and Vincenzo Maria Coronelli (1694).

All three isolarii, published for the first time approximately at regular intervals over the course of a century and a half, belong to the Venetian tradition. Coronelli made his after the sixteenth century and they fall outside the strict scope of this chapter. However, it should be included in the analysis for a better understanding of toponymical discrepancies from a diachronic perspective. These discrepancies exist between the maps made by the different authors, between the toponymical doublets on maps and texts made by the same author, and even between toponyms inscribed on different maps of the same isolario.

In contrast, Rosaccio's *Viaggio...* exhibits significant discrepancies between the toponyms in the text and those depicted on the maps. Furthermore, in his text *Viaggio ...*, the toponym *Carnero* is employed for the Kvarner region, while various maps depict Kvarner in three distinct forms: *Querner*, *QVARNER*, and *QVERRNER*. The ambiguity in the toponymic writing within the texts and maps of the isolarii can be attributed, to some extent, to the coexistence of various traditions in their design and transcription, further corroborated by the differing authorship of the texts and maps, as evidenced in Rosaccio's *Viaggio...*. Namely, the publisher and copperplate engraver Giacomo Franco published both editions of *Viaggio...* and collaborated with Rosaccio as a map engraver. His role can be read from the work itself, specifically from the dedication of *Viaggio's* first edition to Marco Veniero (Rosaccio 1598, 1606). Franco's authorship of multiple maps in Rosaccio's *Viaggio...* is substantiated by his signature on them, as well as by the preponderance of nearly identical maps he published slightly earlier in his work (Franco 1589). Rosaccio's toponyms *Pola* and *Zara* correspond to the forms *Polla* and *Zarra* on maps, which also identify Franco in his cartographic representations (Pettricioli 1965). Concerning the application of toponyms, he committed a significant error in *Viaggio...*, which was similarly evident in the

Table 7.1 Toponymic doublets in various isolarii texts and maps

Geographical feature	*Toponyms in texts*	*Toponyms on maps*
Benedetto Bordone: Isolario (1547)		
Istria (peninsula)	*Listria*	*istria*
Krk (island)	*Vegia*	*uegia*
Cres (island)	*Cherso*	*cherso*
Lošinj (island)	*Oscero*	*oscero*
Rab (island)	*Arbe*	*Arbe*
Pag (island)	*pago*	*Pago*
Trogir (island and town)	*Trau*	*Trau*
Brač (island)	*braccia*	*brazza*
Hvar (island)	*Liezena*	*liezena*
Vis (island)	*Lissa*	*Lisa*
Lastovo (island)	*Augusta*	*lagusta*
Korčula (island and town)	*Curciola*	*curzola*
Mljet (island)	*Melena*	*meleda*
Pelješac (peninsula)	–	*sabiocello; sabioncello*
Giuseppe Rosaccio: *Viaggio da Venetia a Costantinopoli* ... (1598)		
Istria (peninsula)	*Istria*	ISTR IA
Kvarner (gulf, bay)	*Carnero*	Querner, QVARNER; QVERRNER
Rovinj (city)	*Rouigno*	Rouigno, ROVIGNO
Pula (city)	*Pola; Puola*	Pola; POLLA
Cres (island)	*Cherso*	Cherso
Lošinj (island)	*Ossero*	OSERO
Rab (island)	*Arbe*	ARBE
Pag (island and town)	*Pago*	PAGO

(*continued*)

Table 7.1 (continued)

Geographical feature	*Toponyms in texts*	*Toponyms on maps*
Dalmati (region)	*Dalmatia*	DALMACIA
Zadar (city)	*Zara*	ZARRA
Šibenik (city)	*Sebenico*	SIBENICHO
Trogir, town	*Traù*	TRAV
Split (city)	*Spalatro*	SPALATO
Brač (island)	*Brasca*	–
Hvar (island)	*Lesina*	LIESENA
Vis (island)	*Clissa*	–
Korčula (island)	*Curzola*	CRVZOLA, *Curzola*
Mljet (island)	*Melada*	–
Lastovo (island)	–	*Augusta*
Pelješac (peninsula)	–	*Sabionceli*
Dubrovnik (city)	*Ragugi*; *Ragugia*; *Ragusi*	RAGVSI
Vincenzo Maria Coronelli: *Mari, Golfi, Isole, Spiaggie, …* (1694)		
Istria (peninsula)	*Istria*	Istria
Kvarner (gulf, bay)	*Quarnero, Carnero*	QUARNRO; QUARNER,
Poreč (town)	*Parenzo*	PARENZO
Pula (city)	*Pola*	POLA
Rovinj (city)	*Rouigno*	ROUIGNO; ROVIGNO
Cres (island)	*Cherso*	CHERSO, Cherzo
Lošinj (island)	*Osero*, *Ossoro*; *Ossor* (Slav.)	OSERO
Krk (island)	*Veglia*	VEGLIA
Rab (island and town)	*Arbe*; *Rab* (Slav.)	ARBE
Pag (island and town)	*Pago*	PAGO
Dalmatia (region)	*Dalmazia*	DALMAZIA

Zadar (city)	*Zara*	ZARA
Šibenik (city)	*Sebenico*	SEBENICO
Trogir (island and town)	*Traù*	TRAU
Split (city)	*Spalato*	SPALATO
Brač (island)	*Brazza*	BRAZZA
Hvar (island)	*Liesina*	LIESINA
Korčula (island)	*Curzola –*	CURZOLA
Pelješac (peninsula)	*Sabioncello*	SABBIONCELLO
Mljet (island)	*Meleda*	–
Lastovo (island)	*Lagosta*	–
Vis (island)	–	LISSA
Dubrovnik (city)	*Ragusi*; *Dabrounich* (Illir.)	RAGUSI

organisation of the maps within his work. Rosaccio asserts that the island of Vis (*Clissa*) is situated opposite the island of Hvar (*Lesina*), but further towards the sea (*Golfo*) (Rosaccio 1598, 18v). However, the toponym *Clissa* was the name for the fortress Klis located in the mainland, north-west of the city of Split, whereas the island of Vis is most often referred to as *Lissa*. Likely influenced by the resemblance of these toponyms, Rosaccio placed the map of Klis in the section discussing the island of Vis. The maps are numbered in the upper left corner, with the map of Split labelled 12, the map of Hvar labelled 14, and the map of Klis positioned after Hvar, despite being numbered 13. The map numbers in Rosaccio's *Viaggio…* correspond exactly with those on the identical maps of the aforementioned work by Franco (Kozličić 1995). Following the publication of *Viaggio…*, Rosaccio acknowledged his oversight and, in the inaugural edition, subsequent to the Preface, noted in the errata corrige: *a carte 18. facciata seconda verso primo doue dice Clissa vuol dir Lissa* (English translation: *on the map 18, second page, first side, where it says Clissa means Lissa*) (Rosaccio 1598, 1606).

The smallest discrepancies between the toponyms in the text and the maps are found in Coronelli's work, which is justifiable. The standardisation of the language was already established, and over time, there was a notable convergence of the two distinct traditions—inscribing toponyms in texts and on maps—culminating in more uniform forms in both domains.

4 Conclusions

The authors of the isolarii encountered various limitations and sought appropriate solutions. A degree of subjectivity in the creation of cartographic representations is inevitable. It relied not only on the prevailing circumstances but also on the desired outcome of the authors. Prominent features were emphasised at the expense of lesser-known elements. Owing to the substantial demand for such works, numerous authors frequently replicated the creations of their predecessors. The geographical understanding of the coast has not attained a level that permits equitable access to all its areas. It was particularly challenging to depict the regions being mapped for the first time, as the author could not utilise existing maps and had to depend on limited information regarding the dimensions and characteristics of the area. The information regarding obscure areas or coastal segments would be conveyed in a general manner or entirely excluded.

The disconnection among various sectors of human endeavour and inadequate information accessibility hinder the emergence of new quality. Ultimately, it was required that the work possess not only utility but also intrinsic artistic value, resulting in instances where greater emphasis was placed on its decoration and embellishment than on its substance. This significantly affected the credibility of the map content.

References

Astengo C (2007) The renaissance chart tradition in Mediterranean. In: Woodward D (ed) The history of Cartography, vol. 3, Cartography in the European Renaissance. The University of Chicago Press, pp 174–263

Babić I (1984) Prostor između Splita i Trogira. Kulturnohistorijska studija, Muzej grada Trogira

Bliss DP (2013) A history of wood engraving. Skyhorse Publishing

Bordone B (1547) Isolario Nel qual si ragiona di tutte l'Isole del mondo, con li lor nomi antichi, et moderni, historie, fauole, et modi del loro viuere, et in qual parte del mare stanno, et in qual parallelo et clima giacono. M. Federico Toresano

Büsching, AF (1778) Nuova geografia di Ant. Federico Busching, tradotta in lingua toscana ed accresciuta di note dall'ab. Gaudioso Jagemann, vol. XXIII, Tomo vigesimo terzo, che comprende la continuazione della Marca Trevigiana, il Bassanese, il Feltrino, il Friuli, l'Istria, la Dalmazia, ed il Levante. Presso Antonio Zatta

Camocio GF (1571) Isole famose porti, fortezze, e terre maritime sottoposte alla Ser.ma Sig.ria di Venetia, ad altri Principi Christiani, et al Sig.or Turco, nouamente poste in luce, Libreria del segno di S. Marco

Campbell T (1987) Portolan charts from the late thirteenth century to 1500. In Harley JB, D. Woodward D (eds) The history of Cartography, vol. 1, Cartography in prehistoric, ancient and medieval Europe and the Mediterranean. The University of Chicago Press, pp 371–463

Cartwright W (2010) Mapping other (geographical) realities. In: Kriz K, Cartwright W, Hurni L (eds) Mapping different geographies. Springer, pp 11–35. https://doi.org/10.1007/978-3-642-15537-6

Coronelli VM (1694) Mari, Golfi, Isole, Spiaggie, Porti, Citta, Fortezze, Ed altri Luoghi Dell'Istria, Quarner, Dalmazia Albania Epiro, e Livadia, Delineati, e Descritti Dal P. Generale Coroneli, vol. III-IV

Faričić J, Simičić Z (2005) Kornati na starim geografskim i pomorskim kartama/ Kornati Islands on Old Geographic Maps and Charts. Kartografija i geoinformacije 4(4):78–110. ISSN 1333-896X; ISSN 1848-0713

Faričić J, Marelić T, Levačić P, Šinko-Depierris Đ (2020) Hrvatski otoci na kartama u Velikom izolaru i peljaru Andréa Theveta/The Croatian Islands on

maps in André Thevet's Le Grand Insulaire et Pilotage. Hrvatski geografski glasnik 82(2):39–79. https://doi.org/10.21861/HGG.2020.82.02.02
Faričić J, Lončar I (2024) Jadranski otoci u Općem izolaru svih otoka svijeta Alonsa de Santa Cruza. Radovi Zavoda za povijesne znanosti HAZU u Zadru 66:41–81. https://doi.org/10.21857/mzvkptxnx9
Fortis A (1774) Viaggio in Dalmazia. Presso Alvise Milocco
Formaleoni, V (1786) Topografia veneta ovvero Descrizione dello stato veneto. Secondo le più autentiche relazioni e descrizioni delle provincie particolari dello stato marittimo, e di terra ferma, vol I, presso Giammaria Bassaglia
Franco G (1589) Raccolta de'Disegni di tutte le città et Fortezze di Mare che sono nel viaggio Da Venezia a Constantinopoli
Harley JB (2001) In: Lexton P (ed) The new nature of maps: essays in the history of cartography. The John Hopkins University Press
Hind AM (1923) A history of engraving & etching: from the 15th century to the 1914. Houghton Mifflin Company
Kljajić I, Razum M (2016) Izolari Antonija Milla s kartama hrvatskih otoka, Kartografija i geoinformacije 15(25):153–199. ISSN 1333-896X; ISSN 1848-0713
Kozličić M (1990) Istočni Jadran u Geografiji Klaudija Ptolemeja. Latina et Graeca
Kozličić M (1995) Monumenta cartographica maris Adriatici Croatici. Kartografski spomenici hrvatskog Jadrana, AGM
Lago L, Rossit C (1981) Descriptio Histriae. La penisola istriana in alcuni momenti significativi della sua tradizione cartografica sino a tutto il secolo XVIII. Unione degli Italiani dell'Istria e di Fiume, Università popolare di Trieste
Marelić T (2022) The origin of distance and bearing navigational data contained in portolani for the Adriatic Sea basin. Int J Cartography 1–23. https://doi.org/10.1080/23729333.2022.2047402
Novak D, Mlinarić D, Lapaine M (2005) A comparative analysis of the 16th century Ottoman mapping of the Croatian Coast and Islands/Usporedbena studija osmanskog kartografiranja hrvatske obale i otoka u 16. stoljeću. Kartografija i geoinformacije 4(4):78–110. ISSN 1333-896X; ISSN 1848-0713
Pavić M (2003) Plovidbene rute srednjim i južnim Jadranom u izolaru Giuseppea Rosaccia. Radovi Zavoda za povijesne znanosti HAZU u Zadru 45:153–199
Pavić M (2012) Prostorna percepcija istočnojadranskih otoka u očima stranih putnika od XVI. do XVIII. stoljeća. Radovi Zavoda za povijesne znanosti HAZU u Zadru 54:47–68
Pavić M (2013) Otočnici (izolari) – svjedočanstva o europskom i mediteranskom identitetu Hrvatske/Isolarii – testimonies of European and Mediterranean identity of Croatia. Miscellanea Hadriatica et Mediterranea 1:199–213. ISSN 1849-0670, ISSN 2718-1170
Pettricioli I (1965) Ikonografija Zadra do pada Mletačke Republike. Radovi Instituta JAZU u Zadru 11–12:493–526

Prijatelj K (1951) Spomenici Splita i okolice, Turist

Rosaccio G (1606) Viaggio da Venetia a Costantinopoli Per Mare, e per Terra, insieme quello di Terra Santa. Da Gioseppe Rosaccio con brevita descritto.

Rosaccio G (1992) Viaggio da Venetia, a Costantinopoli Per Mare, e per Terra, insieme quello di Terra Santa. Da Gioseppe Rosaccio con brevita descritto (M de Grassi, ed). Edizioni della Laguna S. R. L. (Original work published in 1598)

Tolias G (2007) Isolarii, fifteenth to seventeenth century. In: Woodward D (ed) The history of cartography, vol 3, Cartography in the European Renaissance. The University of Chicago Press, pp 263–484

CHAPTER 8

Mapping the Adriatic: Ottoman Nautical Charts Before the Treaty of Karlowitz

Ahmet Üstüner and Okan Büyüktapu

1 Introduction

The Ottoman Empire's cartographic practices during the fifteenth, sixteenth, and seventeenth centuries represent an intersection of Eastern and Western traditions, particularly exemplified in their mapping of the Adriatic Sea region. This synthesis emerged at a crucial period when the Ottoman Empire was expanding its territorial reach and maritime influence, necessitating increasingly sophisticated geographical knowledge and navigational tools. The development of Ottoman cartography can be traced through three distinct phases: initial engagement with European cartography during Mehmed II's reign (1444–1446; 1451–1481), a golden age of maritime innovation in the sixteenth century, and a transformative period of translation movements in the seventeenth century. Previous works on geographical knowledge in the early modern Ottoman Empire have

A. Üstüner (✉)
Research Institute of Turkology, Istanbul University, Istanbul, Türkiye
e-mail: ahmetustuner@istanbul.edu.tr

O. Büyüktapu
Faculty of Letters, History Department, Ege University, Izmir, Türkiye
e-mail: okan.buyuktapu@ege.edu.tr

J. Faričić, T. Marelić (eds.), *Early Modern Nautical Charts of the Adriatic Sea*, https://doi.org/10.1007/978-981-96-8046-7_8

established the foundational role of Mehmed II in initiating scholarly projects that incorporated Western cartographic knowledge. However, there remains significant debate about the extent and nature of cultural exchange between Ottoman and European cartographic traditions during this period. This study aims to examine how Ottoman cartographers integrated Islamic scholarly foundations with European cartographic advancements to meet their empire's expanding territorial and maritime needs. Of particular interest is the evolution from primarily textual descriptions to detailed nautical charts, exemplified in seminal works such as Piri Reis's (147?–1553) *Kitâb-ı Bahriye* and Kâtib Çelebi's (1609–1657) *Cihânnümâ*. The Adriatic Sea region serves as an ideal case study for this research, as it represented a critical frontier where Ottoman, Venetian, and Habsburg interests converged. By analyzing the development of Ottoman nautical charts of the Adriatic before the Treaty of Karlowitz, this research seeks to illuminate broader patterns of knowledge transfer, cultural exchange, and the relationship between cartographic innovation and imperial expansion in the early modern Mediterranean world. Understanding this synthesis of traditions not only provides insights into Ottoman cartographic practices but also offers valuable perspectives on the broader dynamics of cross-cultural scientific exchange in the early modern period.

2 Ottoman Cartographic Tradition

The Ottoman Empire's geographical knowledge and cartographic tradition evolved over several centuries, reflecting the empire's unique position at the crossroads of Eastern and Western civilizations. This evolution was characterized by a synthesis of Islamic scholarly foundations, European innovations, and the practical needs of an expanding empire. In the early days of the Ottoman Empire, maps were scarce, with the earliest surviving examples dating only from the mid-fifteenth century. Prior to the sixteenth century, Ottoman scholars concentrated primarily on studying and translating Eastern works. However, this began to change dramatically during the reign of Sultan Mehmed II, who demonstrated a remarkable interest in geography and cartography that would prove pivotal in Ottoman history. After conquering Constantinople, Mehmed II initiated several scholarly projects, particularly showing great interest in Western cartographic knowledge. His vision for the empire's future and recognition of geography's importance in state expansion led him to commission various geographical works and maps (Brotton 1997: 87–119). A significant development came when Mehmed II commissioned George Amiroutzes (Georgios Amirutzes) of Trebizond to translate Ptolemy's *Geographiae*. Amiroutzes, who likely worked with his son Mehmed on this project, completed Arabic translations preserved in the Süleymaniye Library as manuscripts Ayasofya 2610 and 2596 (Al-Khaffaf 2023: 351–360). The Sultan specifically requested a world map that would combine Ptolemy's regional maps into a single comprehensive representation, showing his particular interest in practical cartographic applications (Benabbes 2020: 417–427). This period marked the beginning of Ottoman engagement with European cartographic advancements, though the momentum would unfortunately diminish after Mehmed II's death. During his reign, the Topkapı Palace Library was enriched with various geographical works, including multiple copies of Ptolemy's work in Greek, Latin, and Arabic translations (Üstüner and Çinar 2024: 453–460).

The sixteenth century marked a significant turning point for Ottoman cartography, coinciding with the empire's territorial expansion. During this period, Ottoman scholars began producing their own original works, covering a wide range of topics. These new studies encompassed various fields of geographical knowledge. Piri Reis's portolan charts marked the beginning of an era that produced significant contributions to maritime navigation, including nautical charts and treatises on maritime affairs. Additionally, this era saw the creation of detailed information about road networks, travel accounts, and comprehensive studies of specific cities. This shift from primarily translating Eastern works to creating original material reflected the empire's growing need for geographical information to support its expansion and administration.

The development of Ottoman cartographic institutions and practices continued to evolve through the late sixteenth and early seventeenth centuries. A crucial source of information about this evolution comes from Evliya Çelebi (1611–1684?), whose extensive travelogue provides valuable insights into the professional organization of cartography in Istanbul. His accounts detail the existence of a specialized guild of cartographers who employed multilingual mapmakers proficient in languages including Latin, and maintained a commercial enterprise selling their works to sailors and other clients. Çelebi's observations about the involvement of painters (nakkaş) in mapmaking suggest an intriguing intersection between artistic and cartographic practices (Evliya Çelebi 2006: 256). However, it remains unclear whether maps were produced by trained cartographers or by artists commissioned for the task, adding another layer of complexity to our understanding of Ottoman cartographic practices.

The study of premodern Ottoman cartography faces several challenges. The vast scope of the Ottoman Empire's cultural heritage, combined with the scattered nature of its records, makes it difficult to trace comprehensive cartographic works. Inadequate cataloguing of Ottoman cartographic records further hinders research efforts. As cataloguing improves, new examples of Ottoman cartography may come to light, potentially reshaping our understanding of this rich tradition (Karamustafa 1992: 207).

In the seventeenth century, Ottoman geographical studies underwent significant transformation through European influences, primarily via translations. This change was largely driven by Kâtib Çelebi, whose works marked a turning point in Ottoman geography and cartography. The first major European cartographic work to be translated was Mercator's (1512–1594) *Atlas Minor* in 1655, which Çelebi translated with the help of Sheikh Mehmed Ihlasi, titling it *Levâmi'u'n -Nûr fi Zulmet-i Atlas Minor* (Flashes of Light in the Darkness of the *Atlas Minor*) (for the full transcription of the mentioned manuscript, see Üstüner 2017; Mercator and Hondius 1621). Çelebi's most significant contribution was his famous work *Cihânnümâ* (Mirror of the World), which he attempted to write twice. The first version followed classical Islamic geographic traditions, but after translating *Atlas Minor*, he abandoned this approach for a more European style in his second version. Although both versions remained incomplete, they were widely copied and translated. His *Levâmi'u'n-Nûr* was particularly notable for its original commentary-translation and reinterpreted maps from *Atlas Minor*, featuring drawings that approached contemporary cartographic understanding. The integration of European cartographic methods was not merely a matter of translation; the maps were adapted to Ottoman Turkish and Ottoman cartographic style, effectively introducing European cartography to the Ottoman Empire in a culturally accessible form. This period marked a significant transition where Ottoman cartography began to be practiced under the influence of European methods, while still maintaining aspects of classical methodology (Kâtib Çelebi 2017: 47–67).

Çelebi's other geographical works included *Müntehab-ı Bahriye-i Piri Re'îs bin el-Hâc Mehmed Re'îs Hemşîrezâde-i Kemâl Re'îs* (Selected Works from the Navy of Piri Reis, son of Haji Mehmed Reis, nephew of Kemal Reis) (Kâtib Çelebi 1645/1646) and *Tuhfetü'l-Kibâr fi-Esfâri'l-Bihâr* (Gift to the Great Ones on Naval Campaigns) (Kâtib Çelebi 1657; Kâtib

Çelebi 2018: 30–66). Both works were written in response to the Cretan Wars, with *Tuhfetü'l-Kibâr* being completed in 1656. These works combined traditional Ottoman naval knowledge with newer cartographic approaches, demonstrating the gradual synthesis of Eastern and Western cartographic traditions in Ottoman scholarship. This integration of European cartographic knowledge alongside traditional Ottoman methods marked the beginning of modern systematic geography and cartography in the Ottoman Empire, setting the stage for future developments in the field.

Hezârfen Hüseyin Efendi (d. 1691) is seen as the second author in the Ottoman scientific world to refer to Western works. Working alongside Ali Ufki Bey (1610–1675), the translator of the Imperial Council, Hüseyin Efendi, who was interested in Western works with his knowledge of Latin and Greek, was in contact with famous orientalists of his time such as Antoine Galland (d. 1715) and Dimitrie Cantemir (d. 1723). In his work Tenkihü't-Tevârihü'l-Mülûk (*Revision of the Histories of Kings*) on general history, he made use of both Eastern and Western sources. The work is rich in geographical content along with historical and ceremonial information. The work, which also benefited from Kâtib Çelebi's Cihânnümâ, conveyed information about China, (Machin, Khitay, Khotan) islands in the China and Indian Seas, and the discovery of America. It also focused on latitude-longitude calculations and mile and parasang calculations. Partial sections of the work have been translated into German, and various studies have been conducted on it (Özdemir 2007; Bekar 2011; Karaçay Türkal 2014: 321–330; Nas 2019).

Another important figure in this period was Ebubekir bin Behram ed-Dımeşkî Efendi (d. 1691), who emerged alongside Hezârfen Hüseyin Efendi in the second half of the seventeenth century. Together, they continued the systematic integration of Western cartographic knowledge that Kâtib Çelebi had initiated, further advancing the synthesis of Eastern and Western cartographic traditions in Ottoman scholarship. Ebubekir Efendi's most significant contribution was his translation of Joan Blaeu's 11-volume *Atlas Maior* (1662), commissioned by Sultan Mehmed IV (1648–1687). This monumental work, undertaken between 1675 and 1685, represented the culmination of Ottoman engagement with European cartography. The nine-volume translation, preserved in the Topkapı Palace Museum Library (Baghdad section, numbers 325–333), covered a comprehensive range of geographical knowledge, from astronomical concepts to detailed regional descriptions. Notably, it included one of the first Ottoman discussions of Copernicus's solar system, marking a significant step in the integration of European scientific knowledge. The work's structure reflected both European and Ottoman interests, covering territories from Sweden to the New World, with particular attention to Anatolia. Ebubekir Efendi later produced *İhtisar-ı Atlas Mayor* (The Epitome of Atlas Major), a summary version with his own interpretations and additions, demonstrating how Ottoman scholars adapted and reinterpreted European cartographic knowledge to suit their own cultural and intellectual frameworks (Üstüner 2019: 44–45).

3 Ottoman Nautical Charts of the Adriatic Sea and Reproduction of the Adriatic and Its Coasts on Ottoman Maps

The Adriatic Sea has been a focus of cartographic interest since the early fourteenth century, which underscored its pivotal role as a maritime nexus. The development of nautical charts for this region reflects the evolving understanding of geography and the interplay of political, economic, and cultural forces shaping the Mediterranean. The history of Adriatic cartography begins with Pietro Vesconte's 1318 chart, a pioneering work that established a tradition of mapping this key maritime region (Faričić et al. 2023: 24–25). During the sixteenth century, cartographers, particularly from Italian and Venetian traditions, produced nautical charts and maps of the Adriatic coastline with varying degrees of detail and accuracy, motivated by the region's strategic importance in Venetian-Ottoman conflicts. While there was continuous interest in mapping the area, early maps often evolved gradually through incremental improvements rather than major innovations. Cartographers frequently based their work on existing templates, with subsequent mapmakers copying and modifying earlier works—as evidenced by how Merian's maps from a century later clearly borrowed from Rota's earlier works. The most significant improvements came from local cartographers like Martin Rota Kolunić and Natale Bonifacio, who could incorporate firsthand knowledge of the region into their maps. These local experts contributed to better accuracy in depicting geographical features, fortifications, and settlements. However, even with these contributions, cartographic advancement remained a gradual process, with maps often being refined and updated versions of previous works rather than completely new representations (Mlinarić 2011: 72–91).

The Ottoman Empire's cartographic representation of the Adriatic Sea emerged from a rich confluence of Eastern and Western traditions, particularly influenced by the sophisticated Mallorcan cartographic school and classical Islamic geography. This synthesis reflected the Empire's unique position at the crossroads of civilizations and its pragmatic approach to adopting useful knowledge from various sources. The Mallorcan cartographic tradition, which reached its zenith in the fourteenth century through works like Angelino Dulcert's 1339 charts (Dulcert, 1339) and Abraham Cresques's *Catalan Atlas* (Cresques, 1375), established fundamental principles that would later influence Ottoman cartography. This influence manifested through early Islamic cartographers such as Ahmed the Tangier (1413), Ibrahim al-Mursi (1461), and Al-Hac Ebu'l-Hasan (1560?), who adapted the Mallorcan tradition's linear projection techniques and toponym conventions. This cartographic lineage would later inform Ottoman approaches to mapping the Adriatic region. To understand this evolution in practice, we can examine several key cartographic works that exemplify the Ottoman synthesis of Eastern and Western traditions. One of the earliest and most significant examples is Ahmed the Tangier's *Mediterranean Portolan Chart*, which demonstrates this sophisticated integration.

Tangier's portolan chart includes detailed depictions of the European and African coasts of the Atlantic Ocean, British Isles, and Black Sea. It covers the entire Mediterranean and includes a diagram of the lunar mansions (Figs. 8.1 and 8.2). A red arabesque border and brilliant pigments are used, while there are no flags and pennons, elaborate wind roses, or vignettes. Place names are written in red and black in Maghrib script (Ahmed the Tangier, 1413).

Ibrahim al-Mursi's nautical chart (Fig. 8.3), drawn on the gazelle hide, covers the entire Mediterranean and Black Sea, with a calendar in the neck of the parchment at its eastern end. Its margins are decorated in an arabesque red-and-white plaited border, and the place names are written in Maghribî. Although the main sources of the chart were Western, it should be pointed out that considerable original additions to the toponymy of the Islamic territories have been made (Ibrahim al-Mursi, 1461). This fusion of Western and Islamic cartographic traditions would continue to evolve in the following century, as exemplified by another groundbreaking work.

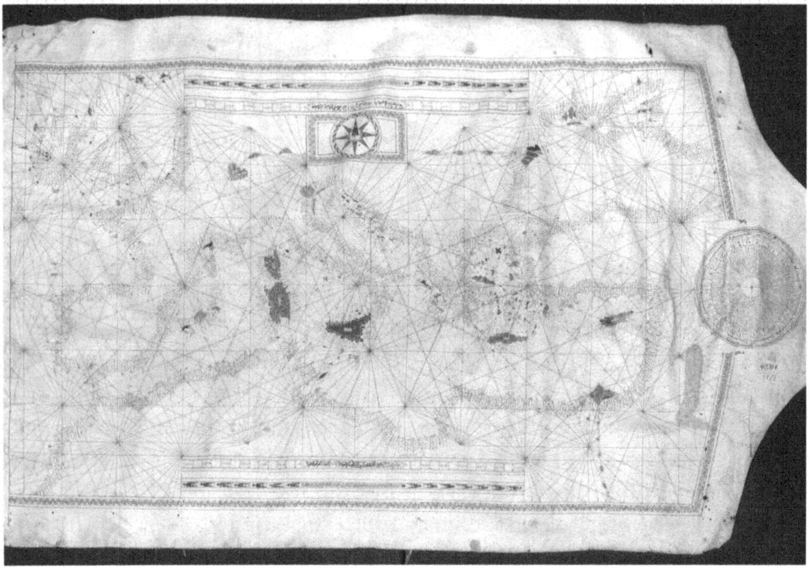

Fig. 8.1 Ahmed the Tangier, Mediterranean portolan chart from 1413 (Source: Topkapı Palace Museum Library, Hazine Collection, Manuscript No. 1823)

4 The Ptolemaic Revolution in Ottoman Mapping

While early Ottoman cartographers successfully merged various traditions, it was during Sultan Mehmed II's reign (1444–1446; 1451–1481) that Ottoman cartography underwent its most significant transformation. This pivotal period was marked by the Sultan's initiative to translate Ptolemy's *Geographiae* (known in Ottoman as Batlamyus, *Kitâbü'l-Coğrafya fi'l-Ma'mûre mine'l-Arz*) into Ottoman scholarship. This groundbreaking work transformed geographical science through several key innovations. First, it introduced a scientific methodology to geography, incorporating mathematical concepts that would prove fundamental to future cartographic developments. Ptolemy's innovative use of latitude and longitude coordinates for mapping major centers of the known world built upon Hipparchus's earlier proposals from three centuries before. Furthermore, his division of geographical study into two distinct categories—geography and topography—represented a significant advancement in the field's organization and understanding. The influence of Ptolemy's work

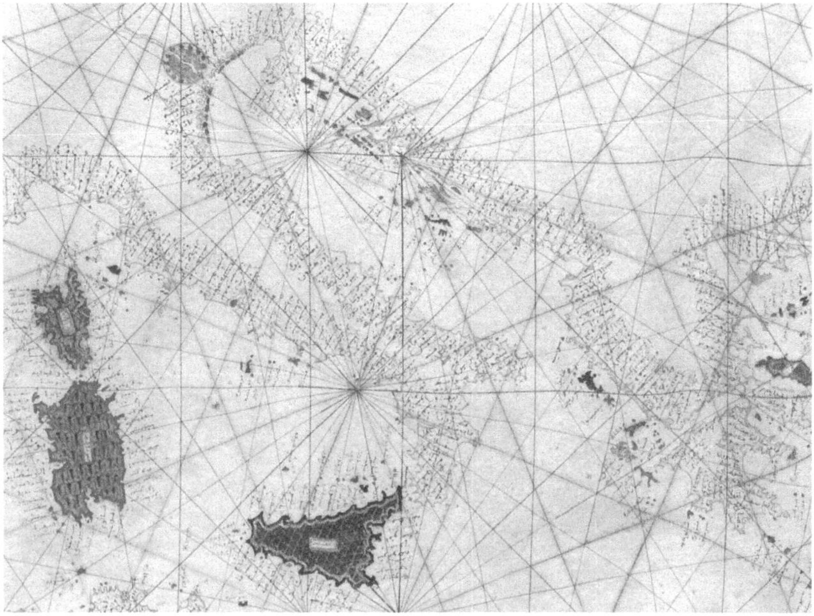

Fig. 8.2 Details of the Adriatic Section of Ahmed the Tangier's portolan chart from 1413 (Source: Topkapı Palace Museum Library, Hazine Collection, Manuscript No. 1823)

extended far beyond its initial creation. While it drew heavily from Marinos of Tyre's earlier contributions (though the extent of this influence remains unclear due to limited surviving records), the text's journey through various translations reflects its enduring importance in both Islamic and Ottoman scholarship. The initial Arabic translation by philosopher al-Kindī proved inadequate until Thābit ibn Qurra produced a more accurate version, setting the stage for Sultan Mehmed II's later commissioned translation that would directly influence how Ottoman cartographers approached the mapping of regions like the Adriatic. Sultan Mehmed II's commissioning of a new Greek-to-Arabic translation in the fifteenth century, now housed in the Süleymaniye Library (Ayasofya, no. 2596 and 2610), demonstrates the Ottoman Empire's commitment to preserving and advancing geographical knowledge (Fig. 8.4). This was followed by the first Latin edition in Vincenza (1475), a subsequent edition in Bologna (1482), and the first Greek edition in Basel (1533). A particularly intriguing

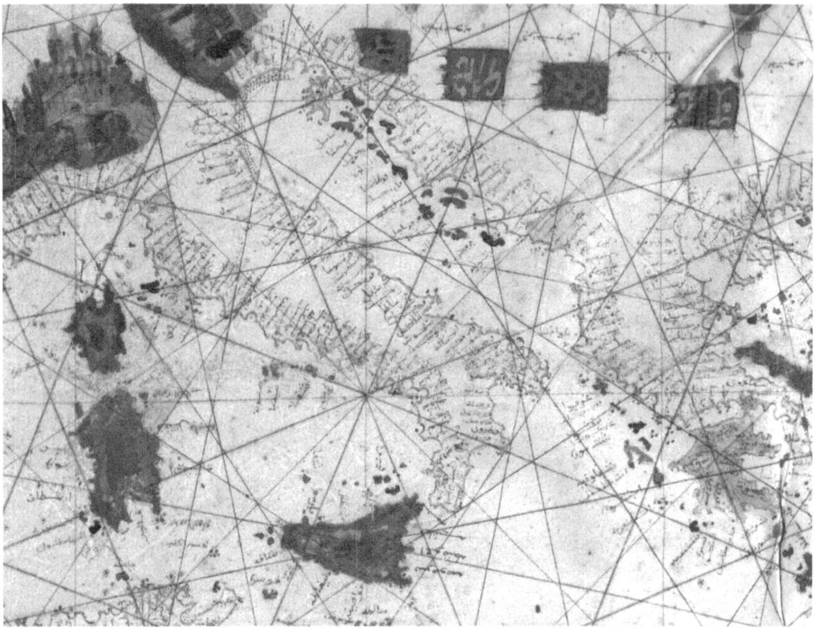

Fig. 8.3 Details of the Adriatic Section of Ibrahim al-Mursi's Nautical Chart from 1461 (Source: Istanbul Naval Museum, Manuscript No. 882)

historical controversy surrounds the maps accompanying the work. While R. Taton argued that these maps were Byzantine additions from the thirteenth and fourteenth centuries, Fuat Sezgin's research revealed their true origins among Islamic geographers during Caliph al-Ma'mun's reign in the ninth century. According to Sezgin, these Islamic maps were later incorporated into the Greek text by the Byzantine scholar Planudes, who mistakenly attributed them to Ptolemy (Sezgin 1987: 7–8). This finding is further supported by the late discovery of Greek manuscripts of Ptolemy's works, whose authenticity continues to be debated in contemporary scholarship (Aydın and Aydın 1992: 198–199).

However, the translation did reveal how Ottoman intellectuals interpreted the Greek text and evaluated the place names. At this point, the Ottoman descriptions of toponyms in Ptolemy's *Geographiae*, "Halîc-i Az(d)ri (the Gulf of Adriatic), Cezâ'ir-i Dıravmizyat (Dalmatian Islands), Ba'zü'l-vilâyetü'l-İlirrisi (Some part of Illyria),

Fig. 8.4 Ptolemy's *Kitâbü'l-Coğrâfya fi'l-Maᶜmûre mine'l-Arz* (Source: Presidency of Manuscript Institution of Turkey, Süleymaniye Manuscript Library, Ayasofya Collection, Manuscript No. 2610, folio 19)

Ba'zü'l-vilâyetü'l-Banoniyetü't-tahtâniye (Some part of Lower Pannonia), Ba'zü'n-nehr-i zâr (Some part of River Sava), e'n-nehrü Sâvi (Sava River), Ba'zü'l-vilâyetü'l-Banoniyetü'l-fevkâniye (Some part of Upper Pannonia), Ba'zü'l-vilâyetü'l-Vendelikiyyetü (Some part of Venetian Territory), Mesâkin-i Karnin (Settlements of Carnia), Mesâkin-i Venedik (Settlements of Venice), Nehr-i Etran (Etran River), Triste (Trieste), Parendiye (Parentium/modern Poreč), El-İstiriye (Istria), Pola (Pula)," are the notable names at first sight along the eastern Adriatic coasts. The Adriatic is again depicted in general outline in the Mediterranean-centered ancient world map drawn in a partially different style in the same copy of Ptolemy's translation (Ptolemy, 1460s, folio 19).

5 Age of Piri Reis and *Kitâb-ı Bahriye*

The eastern Adriatic in the early modern period represented a complex zone of interaction between major European powers, serving both as an arena of conflict and a nexus of cultural exchange. This region, particularly during the sixteenth century, was shaped by the interplay between the Republic of Venice, the Habsburg Empire, and the expanding Ottoman Empire. Each of these powers sought to establish and maintain control over strategic territories and maritime routes. Within this complex political landscape, the cartographic work of Piri Reis, especially his "Kitâb-ı Bahriye" (Book of Sea Lore) (Piri Reis, 1526), stands as a remarkable achievement in maritime cartography. His detailed documentation of the eastern Adriatic coast surpassed contemporary European mapping efforts, particularly those of Venetian cartographers, despite their direct access to the region. The *Kitâb-ı Bahriye*, produced in two versions (1521 and 1526), combined precise nautical charts with comprehensive written descriptions of ports, islands, and navigational features. The innovation of Piri Reis's work lay in its practical utility for navigation and its detailed geographical accuracy. For instance, his charts provided specific information about important features such as the Strait of Osor's drawbridge, the Tower of Orlando in Pula (Tara Pola), and detailed descriptions of harbors like Dubrovnik, including their defensive structures and signaling systems. His work demonstrated remarkable attention to both military and commercial aspects of maritime navigation. Notably, while Venetian cartographers did not produce detailed representations of their Adriatic possessions until the 1570s (with works by Camocio, Pinargenti, and Rosaccio), Piri Reis had already created detailed charts of the region decades earlier. His approach combined traditional Ottoman cartographic elements with European maritime mapping techniques, creating a hybrid style that would influence Ottoman cartography throughout the sixteenth and seventeenth centuries. The superiority of Piri Reis's charts was not matched until Vincenzo Maria Coronelli's work and was only truly surpassed by Giacomo Marieni's *Portolano del Mare Adriatico* in 1830. This achievement is particularly remarkable considering that Piri Reis worked as an outsider, representing a power that was often in conflict with the Republic of Venice, which controlled most of the eastern Adriatic coast. These cartographic works reflect not only geographical knowledge but also the complex political and cultural dynamics of the region. While the Venetians were protective of detailed geographical information about their territories, Piri Reis's work demonstrated the possibility of transcultural exchange of knowledge, even in a period marked by political and military rivalry. His charts and descriptions serve as evidence of the eastern Adriatic's role as a space where different cultural and political traditions could intersect and influence each other, despite the ongoing conflicts and tensions that characterized the region (Mlinarić and Faričić 2024: 381–399).

The earliest surviving copies of Kitâb-ı Bahriye date to 1550, and these versions contain comprehensive mapping and descriptions of the Adriatic region. The Ottoman geographic tradition is particularly evident in the detailed place name descriptions throughout the text. The following place names and nomenclature are derived from figures above:

Zikr-i eşkâl-i şehr-i Venedik (description of Venice), boğazda evler (houses in the channel), kal'a (fortress), kastel (castle), ada kal'asıdur (the island castle), Venedik boğazda evler... (houses in the channel of Venice), manastırlık (monastery area), şişe âlâtı bunda işlenür (glassware is manufactured here), köpridür (this is a bridge), kilisa (church) (Fig. 8.5a).

"Zikr-i eşkâl-i Dubrovenedik (Description of the features of Dubrovnik), bu. cezîrelerün ekseri Dubrovenedik'e tâbi'dür (Most of these islands are under the sovereignty of Dubrovnik). Tâbi'-i Dubrovenedik (sovereignty of Dubrovnik), Cezîre-i Kasekasola (Sušac Islet), Korçıla (Korčula), Liman-ı Dubrovenedik barçalar bunda kışlar (Port of Dubrovnik where barges winter), değirmen burnı ve manastırdur (mill cape and monastery), Kal'a-i Kamışlık (Trsteno?), Dubrovenedik sınurı (Dubrovnik border), Kızıl Kamere (Koločep Island) cezîreleri (Sv. Andrija Islet), Dalmaş (Dalmatia), Cezîre-i Soline (Šolta Island), Cezîre-i Baraça (Brač Island), Cezîre-i Lorda (Šćedro Island)" (Fig. 8.5b).

"Manastır Burnı (Monastery Cape), a'lâ sudur (an excellent water), Ankona şehri (city of Ancona), Boğazda evler ve kal'a, kastel (houses and fortress in the channel, castle), Ada Kal'ası (Island Fortress), Manastırlar (Monasteries), Boğazda evler... (Houses in the channel), Venedik'e akar sudur (it is water flowing to Venice), a'lâ sudur, tersane (an excellent water), Riyale Köprüsi çârşusı bâzârı üzerindedür ve burgunundan(?) kalyon geçer (The Rialto Bridge has markets and bazaars on it, and galleons pass through its arch)" (Fig. 8.6).

In both versions of the work, the Adriatic sections are notably detailed, though the evolution of drawings and toponymy (place names) varies across different copies, reflecting the individual perspectives and approaches of various copyists and mapmakers. While historical evidence suggests that

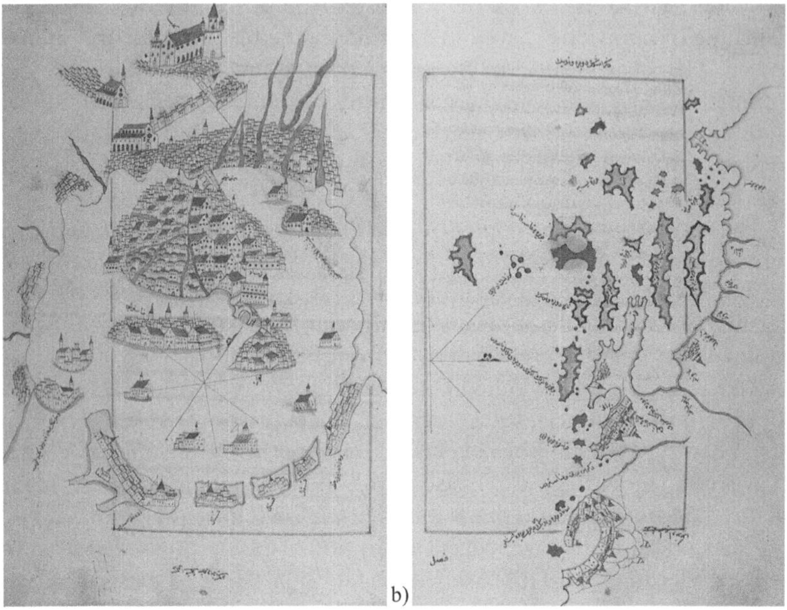

a) b)

Fig. 8.5 Details of Venice (**a**) and the middle and southern parts of the Dalmatian coast and islands (**b**) from Piri Reis's *Kitâb-ı Bahriye* (1521) (Source: Presidency of Manuscript Institution of Turkey, Süleymaniye Manuscript Library, Ali Emiri Manuscript Library, Coğrafya Collection, Manuscript No. 1, folia 78, 80)

charts and chart making were familiar to Ottoman mariners in the late fifteenth and early sixteenth centuries, the *Kitâb-ı Bahriye* established a new standard in maritime cartography, successfully balancing practical navigation needs with artistic representation within the portolan chart tradition. Piri Reis's book goes beyond being a mere geographical treatise, as it is filled with maps and vital information designed to guide sailors. The work serves as a comprehensive navigation manual, containing detailed information about various islands and fortresses, including their ruling authorities. Beyond cartographic elements, the narrative sections of Bahriye function as a complete navigation book, providing sailing routes, port information crucial for mariners, and details about coastal food and freshwater sources. These comprehensive works laid the foundation for subsequent developments in Ottoman maritime cartography, leading to

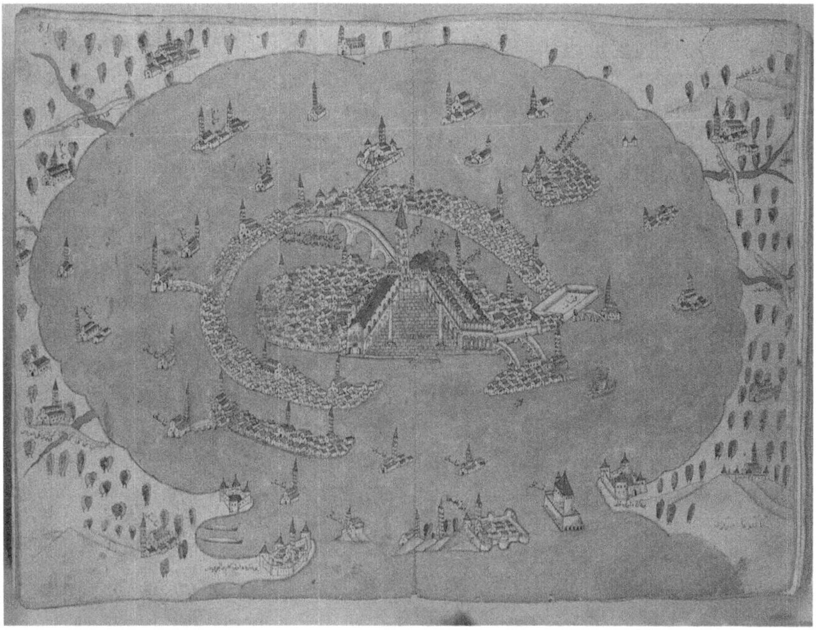

Fig. 8.6 Piri Reis, Venice, in *Kitâb-ı Bahriye* (1521) (Source: Presidency of Manuscript Institution of Turkey, Süleymaniye Manuscript Library, li Emiri Manuscript Library, Coğrafya Collection, Manuscript No. 1, folio 81)

what scholars identify as a distinct transitional phase in the mapping of the Adriatic region.

6 Transitional Period

Following Piri Reis, Ottoman mapping of the Adriatic entered a phase that can be identified as the "Transitional Period" characterized by a synthesis of multiple cartographic traditions. This era is particularly well represented in the portolan atlases of the Walters (*Walters Nautical Atlas*) collection and the remarkable work of *Ali Macar Reis*, along with the *Atlas-ı Hümâyûn* series, all of which flourished under the active patronage of the Ottoman court and its pursuit of geographic knowledge.

Ali Macar Reis was a unique figure who straddled two worlds in the Ottoman Empire of the mid-sixteenth century. He appears in records as early as 1558 as a member of the Guild of Rumi Painters (Cema'ât-i Naḳḳaşân-ı Rûmiyân) employed by the imperial palace, and by 1571 had risen to become a captain of the Sultan's private galleys. His portolan atlas, completed in 1565, synthesized multiple cartographic traditions. This atlas was closely based on Western prototypes. The significance of Ali Macar Reis's work lies in how it demonstrates that Ottoman cartographers actively updated and adapted their maps for practical purposes, combining multiple sources and contemporary intelligence to create strategically useful documents. This challenges earlier assumptions that Ottoman maps were primarily abstract representations of political and religious space rather than practical tools. The timing and content of the atlas suggest it was commissioned specifically to support Ottoman naval ambitions, demonstrating that such maps served practical military and strategic purposes rather than being purely abstract or decorative works. This challenges previous assumptions that Ottoman world maps were primarily expressions of religious and political space rather than practical navigational tools (Casale 2012: 54–62).

A close examination of the toponyms in Ali Macar Reis's atlas further demonstrates this practical approach to cartography. In his detailed representation of the Adriatic coast (Fig. 8.7b), the careful documentation of place names reflects both Ottoman and European naming conventions, illustrating the synthesis of different cartographic traditions. Moving from the eastern shores northward, the atlas records: "Santa, Avlona (Vlora), Dalyan (Daljan), Aylona Suyı (Vlora River), Eski Dıraç (Old Durrës), Arnavud İskenderiyesi (Albanian Alexandria/Durrës), Kotur (Kotor), Şehr-i Dubrovenedik (City of Dubrovnik), Dubrovenedik limanı (Port of Dubrovnik), İsbileto Şehri (City of Split), İncir Burnı (Smokvica), Zadar Şehri (City of Zadar), Bortoro (Portorož), Bona Şehri (City of Nona / Nin), Sen Sancorci (Saint George / Sv. Juraj), İstar Şehri (Capo d' Istria / Koper), San Bero (San Pietro / Sv. Petar), Bosna (Bosnia), Doğan Dağı (Monfalcone), Santandriya (Sant'Andrea / Sv. Andrija), Venedik Şehri (City of Venice)." These place names, combining Ottoman Turkish toponymy with local European names, exemplify the transitional nature of this period in Ottoman cartography. The systematic recording of both maritime features and urban centers reveals how Ottoman cartographers like Ali Macar Reis were creating sophisticated navigational tools that merged multiple cultural and technical traditions to serve the empire's expanding maritime ambitions.

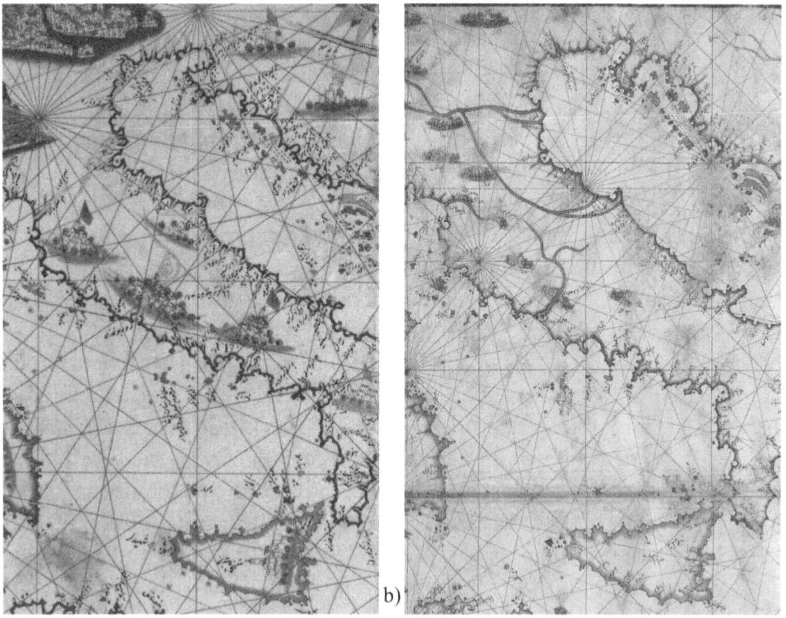

Fig. 8.7 The Adriatic and Tyrrhenian Seas on two portolan charts; (**a**) from the *Walters Nautical Atlas* made in the 1560s (Source: Walters Collection, Manuscript No. W660, folio 7); (**b**) Ali Macar Reis's chart made in the sixteenth century (Source: Topkapı Palace Museum Library, Hazine Collection, Ali Macar Reis Atlası, Manuscript No. 644, folio 4)

7 Translation and Adaptation Period

The mid-sixteenth century marked a significant period of cultural and scientific exchange between Western European and Ottoman cartographic traditions. Apart from the three atlases mentioned above, a prime example of this exchange is the remarkable work created by Hajji Ahmed of Tunisia. His *Cordiform Map* of 1559 stands as a remarkable cartographic achievement that bridges Western European and Ottoman traditions. Created in Venice, the map measured an impressive 115 cm by 111 cm and was constructed from six woodblocks. Though intended for Ottoman audiences, its creation history took an unexpected turn when the woodblocks were confiscated and stored in Venice's Council of Ten archives, only to be rediscovered in the late eighteenth century, leading to the printing of 24 copies, of which 11 survive today. The map's most distinctive feature is its heart shape, following the cordiform tradition popularized by Oronce Fine's 1534 map. It presents the world in two major divisions: the New World in the West and an updated Ptolemaic East. The map is extensively annotated in Ottoman Turkish, with detailed textual commentaries surrounding the main cartographic elements. At the bottom, two celestial hemispheres complement the terrestrial representation, demonstrating the mapmaker's comprehensive approach to world representation. Artistically, the map incorporates traditional Ottoman decorative elements, including a frieze featuring leaf motifs (*yaprak*), stylized bird's-eye view flowers (*penç*), and semi-stylized botanical forms (*hatayi*) patterns. This artistic choice reflects the map's intended audience while demonstrating the sophisticated integration of Western cartographic knowledge with Ottoman artistic traditions. One of the map's most intriguing aspects is its system of correspondence, linking seven great rulers with the seven known planets of the solar system. This feature, combined with the use of Turkish language (which the mapmaker claimed was "the most dominant in the world"), reveals the map's role in both scientific documentation and cultural diplomacy. The significance of Hajji Ahmed's map extends beyond its cartographic value. It represents a crucial moment in cross-cultural exchange between Western European and Ottoman civilizations during the sixteenth century. The map demonstrates how scientific knowledge could transcend cultural and religious boundaries, while simultaneously maintaining distinct artistic and cultural identities (Barthe 2008: 32–44). The map and treatise primarily cover the geography of the Adriatic region. The area served as the Ottoman Empire's maritime gateway to the West and contains essential nautical information (Tunuslu Hacı Ahmed 2019).

While Hajji Ahmed's map represented an individual masterpiece of Ottoman–European synthesis, broader institutional developments were simultaneously reshaping Ottoman cartography. During the second half of the sixteenth century, Ottoman cartographers began systematically incorporating European geographical knowledge into their works, particularly through Italian-style world maps. A particularly significant example of this cultural exchange is *The Muhit Atlas* (Oceanic Atlas), a virtually unknown Ottoman manuscript atlas preserved in the Biblioteca Nazionale in Naples. This atlas provides important evidence of cartographical and geographical knowledge transfer between Italy and the Ottoman Empire. Internal evidence suggests that the prototype for The *Muhit Atlas* was a Venetian set of globe gores, likely associated with Giacomo Gastaldi's 1561 world map.

While the original globe appears to have been lost, the atlas enables a reconstruction of its main features. The atlas consists of 171 plates systematically presenting the earth's surface, starting at the equator and moving through both hemispheres in 15-degree bands. The atlas demonstrates clear Italian influences through its place names and geographical outlines, while also showing Ottoman adaptation through Turkish toponymy and legends. The compiler appears to have been a highly educated Ottoman scholar with reading knowledge of Italian, though not necessarily experienced in cartography. The atlas contains detailed commentaries on historical, ethnographic, and geographical information drawn from both Islamic and European sources. *The Muhit Atlas* (Fig. 8.8) is particularly unique as it seems to have left the Ottoman Empire shortly after its creation in the late sixteenth century, ending up in the possession of the del Tufo family in Naples. This early departure may explain why it remained a unique specimen without influencing broader Ottoman cartographic traditions. The atlas represents an important example of cross-cultural scientific exchange in the Mediterranean world during the early modern period,

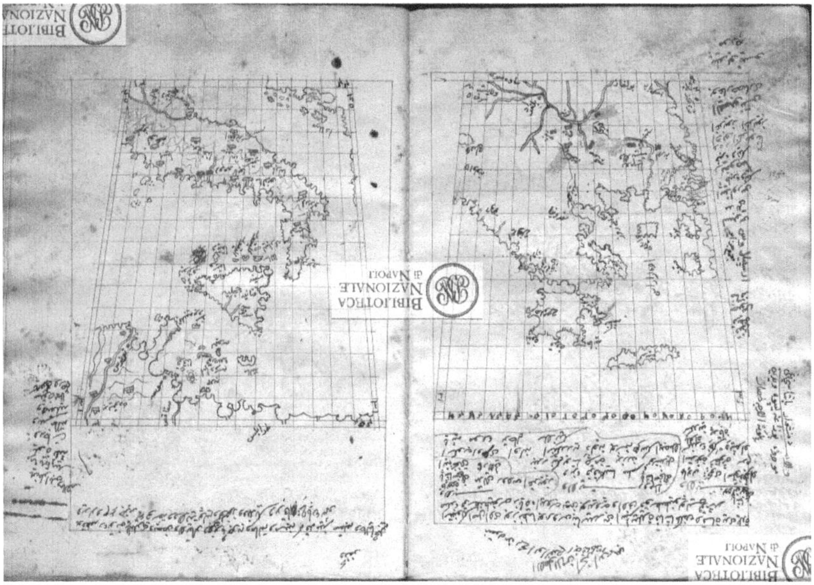

Fig. 8.8 "Adriatic and Italy," *The Muhit Atlas* (Source: Biblioteca Nazionale di Napoli, BNN Ms. III.F.35, folio 99)

while also providing insight into both Ottoman reception of European cartography and previously unknown aspects of Venetian map production (Berardi 2017: 37–51).

The map of the Adriatic region, though partially in draft form, shows notable city markers including Vlora (Avlona), Milan, Venice, and Ancona. While the poor quality of the paper has resulted in a loss of detail, Italy's distinctive boot-like shape nevertheless makes it easy to identify the Adriatic Sea and its surrounding geography.

The important development of the seventeenth century was that the partial interaction of the previous century influenced Ottoman geography in the form of direct translation and adaptation. As stated earlier, the seventeenth century marked a significant transformation in Ottoman mapping of the Adriatic, largely through the works of Kâtib Çelebi. His *Müntehab-ı Bahriye* (Fig. 8.9), inspired by Piri Reis's *Kitâb-ı Bahriye*, and

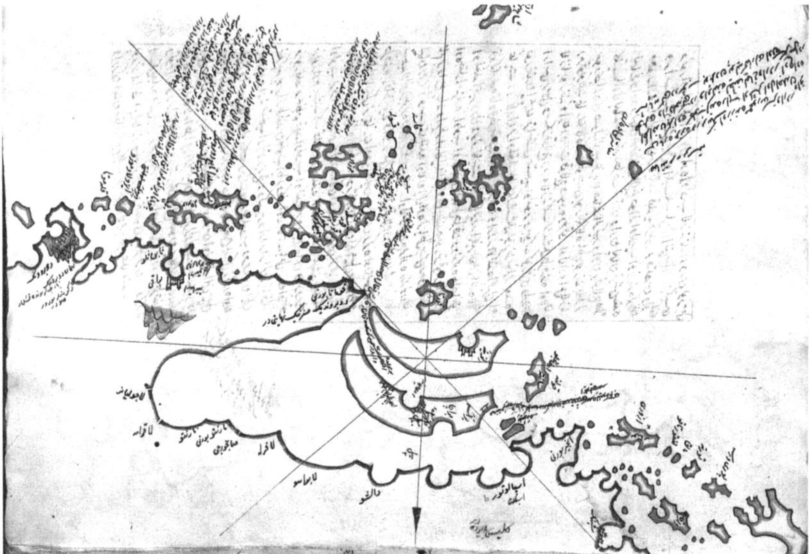

Fig. 8.9 Depiction of the central and southern parts of the Dalmatian coast and islands along with Dubrovik and Korčula in *Müntehab-ı Bahriye* (Source: Presidency of Manuscript Institution of Turkey, Süleymaniye Manuscript Library, Aşir Efendi Collection No. 227, folio 62b)

Tuhfetü'l-Kibâr fi-Esfâri'l-Bihâr provided detailed information about Mediterranean geography, including the Adriatic basin.

Müntehab-ı Bahriye provides extensive detail about the Adriatic coastline, as evidenced in the map's Ottoman Turkish text, which can be translated as follows

"Dubrovenedik (Dubrovnik), Liman-ı Dubrovenedik (port of Dubrovnik), barça gününde kışlar (where barges (barça) winter), değirmenler budur (These are mills), sudur (water-river), İstante, Dubrovenedik'ündür (belongs to Dubrovnik), Komato Burnı (Cape Komata), Dubrovenedik sınırının nihâyetidür (End of the border of Dubrovnik), Narento (Neretva), Marento Burnı (Cape Neretva), Sancorci (St. George/Sv. Juraj), Lakrale, Labraso, Dalşo, İspalotor (Split), İncir Burnı (Smokvica), Poriverse (Mala and Vela Proversa), Torte (Fort Toreta on Kornat Island), Barasa and Baraça (Brazza / Brač Island), Lizeniye (Liesina/Hvar Island), Korsila and Korçila (Korčula Island).

Beş pâre cezîredür. Bunun üzerinde ma'mûr manastır vardur. Etrâfdan beğzâde gelüp anda râhib olmışlar. Ol ecilden vâfir topları var. Kal'a gibidür ve önünde latîf limanı vardur üç ada sepâye (?) gibi düşüp bunlardan Pulye kenârları on sekiz mildür. (*There are five-piece islands. On them stands a prosperous monastery where noblemen's sons from surrounding regions have become monk. For this reason, they have plenty of cannons and it resembles a fortress. In front of it lies a pleasant harbor, with three islands positioned like tripod, from which the Puglia shores are eighteen miles away*).

Biri birine ok yetişir mâbeyni barçalara limandur. Lodos tarafında olan adada bir koy vardur önünde kıbleye bucak yatakdur. (*The distance between them is within arrow shot, providing harbor for barges. On the island) facing the south wind, there is a bay with a south-facing anchorage*).

Çomaba izle de Levanti porto kale ağzındadur iki tarafından limana girilür. Portokale: Uzun adadur ve ... dağları vardur yırakdan yılan gibi görinür. Etrâfı elli mil kadardur. Rum ili kenârına bir mil ve ba'zı yeri iki mil karîbdür. Beş... limanları vardur. Biri keşişleme tarafında Portokale ikinci ahtopot limanı ma'nâsına porto de teşnedil kılipodur barçalar girer. Üçünci Porto de Çiyanodur içinde çıyan gibi bir uzun ada vardur. (*Çomaba island of Levant. Portokale is at the mouth, with harbor entrance from two sides. Portokale: It is a long island with mountains that appear like snake from afar. Its circumference about 50 nautical miles (approximately 93 kilometers). It is one mile (approximately 1.6 kilometers) from the Rumelia shore, and in some places two miles close. It has five harbors: The first is Portokale (Port of Sobra on Mljet Island) on the northeast side, the second is porto de teşnedil (Port of Pomena) meaning octopus harbor where barges can enter, and the third is Porto de Çiyano (Port of Polače) which has a long island inside like a centipede).*"

While the *Müntehab-ı Bahriye* represented traditional Ottoman maritime cartography, Kâtib Çelebi's later work, the *Levâmi'u'n-Nûr fî Zulmet-i Atlas Minor*, demonstrated a significant evolution in Ottoman mapping techniques. This work showcases how Ottoman cartographers began incorporating European cartographic innovations while maintaining their distinct geographical perspective.

Kâtib Çelebi's works represent a synthesis of traditional Islamic geographical knowledge and European cartographic innovations, as is particularly evident in his treatment of the Adriatic coastline and islands. The two maps from the *Levâmi'u'n-Nûr* provide detailed geographical information about the Adriatic region:

Dalmasiya (Dalmatia), Bosna (Bosnia), Kuruvasiya (Croatia), Iskılavoniya (Slavonia), İstiriya (Istria), İtalya (Italy), Marka Ankonitana (March of Ancona), Abruzzo (Abruzzo), Golfo di Venesiya ya'nî Venedik Körfezi (Gulf of Venice), Valone (Vlora/Valona), Durazo (Durrës), Sazene (Sazan Island), Lodrin (Lezhë), Ankiyotari (Antivari/Bar), Kataro (Kotor), Ragozi (Ragusa/Dubrovnik), Konovase (Konavle) (Fig. 8.10a).

Iskılavoniya (Slavonia), Kuruvasiya (Croatia), Bosna (Bosnia), Dalmasiya (Dalmatia), Hungariya'nun cânibi (Hungarian territory/side), San Martin (St Martin/Sv. Martin), Uvvelevar (Bjelovar), Zerdabel (Đurđevac), Sakhon (Suhopolje), Kırahonter (Magyarkanizsa/Kanjiža), Firindav (Virovitica), San Nikalos (St Nicholas/Sv. Nikola), Bakokza (Pakrac), Atona (Athonite/Mount Athos), Dagrapiya (Đakovo), San Corc (St George/Sv. Juraj), Venedik Körfezi (Gulf of Venice), İspelato (Split) (Fig. 8.10b).

In the late seventeenth century, the Ottoman translation movement marked a significant enhancement in the empire's cartographic precision, particularly in its representation of the Adriatic region. A prominent example is Ebubekir bin Behram ed-Dımeşkî's work on translating Joan Blaeu's *Atlas Maior*. This translation effort, which took place between 1675 and 1685, involved adapting European cartographic techniques to Ottoman traditions, while preserving the empire's unique geographical terminologies and perspectives. The project underscored the Adriatic's strategic importance to the Ottoman Empire, reflecting both military and economic priorities. The translated maps, often described as a synthesis of European and Ottoman cartographic traditions, were rendered in

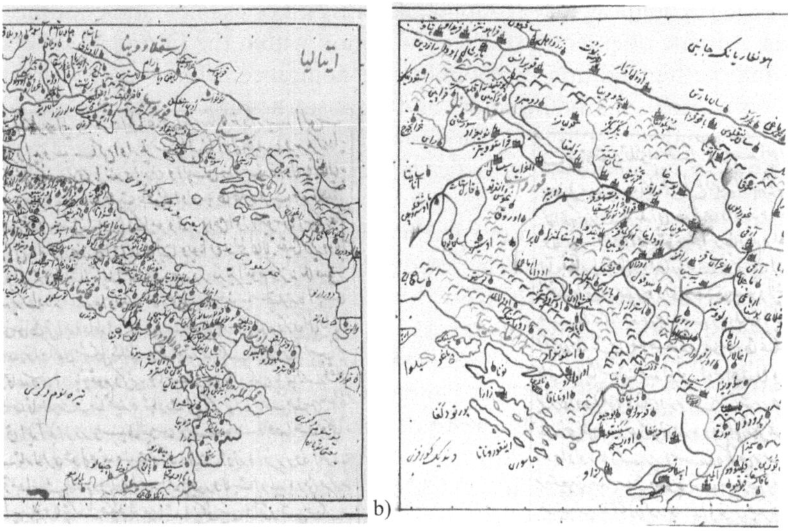

a) b)

Fig. 8.10 The depiction of the Adriatic Sea (**a**) and the depiction of North Dalmatia and the Croatian and Bosnian hinterland (**b**) in *Levâmi'u'n-Nûr fi Zulmet-i Atlas Minor* (Source: Presidency of Manuscript Institution of Turkey, Süleymaniye Manuscript Library, Nuruosmaniye Collection, No. 2998 folia 318a, 359a)

Ottoman Turkish and included detailed summaries. Among the preserved manuscripts, two key versions, Nuruosmaniye 2996 and Türk İslam Eserleri Müzesi 1975, reveal a complex process of cultural adaptation. These works exhibit a combination of Blaeu's scientific precision with elements of Islamic geographical scholarship, showcasing the Ottoman interest in blending imported knowledge with their own intellectual heritage. The integration of Blaeu's methods into Ottoman cartography not only introduced advanced mapping techniques but also fostered a more nuanced understanding of the Adriatic and other vital regions. This cross-cultural exchange demonstrated the Ottomans' ability to reinterpret European cartographic elements while maintaining their own narrative. Additionally, the adaptation of Blaeu's maps into manuscripts like the *Cihânnümâ* by Kâtib Çelebi illustrates how these texts served as a bridge between European innovations and Ottoman geopolitical objectives. The strategic use of maps in these translations extended beyond mere representation. The manuscripts include not only geographical details but also historical, political, and economic narratives. For example, descriptions of the Adriatic often contextualize the region within the Ottoman-Venetian conflicts, highlighting its geopolitical significance. Moreover, the artistic aspects, such as vibrant colors and intricate designs in manuscripts like Nuruosmaniye 2996 and Türk İslam Eserleri Müzesi 1975, reflect the Ottoman emphasis on aesthetics alongside functionality (Brentjes 2012, 171–220).

The map is titled "Sûret-i Italya" (Representation of Italy) and in the map the Dalmatian coast is described in detail. Noteworthy notes and nomenclature from south to north are as follows: "Anti Paso (Antipaxos), Cezîre-i Pakso (Island of Paxos), Kapo Biyanko (Cape Bianco/White Cape), Gudiya (Gouvia), Cezîre-i Körfez (Corfu Island), Körfez (Corfu), Porte Tonon/P. Timon (Port Timon), Oraimus (Sarande), Delonye/C. de Longua (Cape Longa), Boğaz-ı Venedik (Gulf of Venice), Avlona (Vlora), Vilâyet-i Arnavudluk (Province of Albania), Dıraç (Durrës), Lodrin (Lezhë), Liş (Lezha), Ülgün (Ulcinj), Bar (Bar), Budva (Budva), Kotur (Kotor), Nova (Herceg Novi), Dubrovenedik (Dubrovnik), Sancağ-ı Hersek (Herzegovina), Sabiyonçıla (Sabbioncello/Pelješac), Lezine (Hvar), İspolit (Split), Tıravo (Trogir), Arhancilo (Arkanđel Islet), Şebenik (Šibenik), İskardone (Skradin), Zadre (Zadar), Nona (Nin), Lubuşka (Ljubač), Yablaniç (Jablanac), Siyena (Senj), Bukoriçe (Bakar), Fiyomi (Rijeka), Korkula (Korčula), Cezâyir-i Cara (Zadar Islands), Lisso (Vis Island), San Andriya (Sv. Andrija Island), Melizla (Brusnik Islet), Balagoza (Palagruža Islet)" (Fig. 8.11).

The Ottoman cartographic representation of the Adriatic coastline exemplifies a sophisticated synthesis of Western European and Islamic traditions that evolved over several centuries. The comparative analysis of toponyms reveals two distinct but complementary sources of geographical knowledge. First, Ottoman cartographers consciously incorporated Venetian nautical knowledge, as evidenced in works like Ali Macar Reis's 1565 atlas and Piri Reis's *Kitâb-ı Bahriye*, where precise navigational features are documented using terminology derived from Venetian charts. This is particularly apparent in their detailed documentation of strategic maritime features like Premantura (Bolmantor/Polimantor), the Proversa strait (Poriverse), and Smokvica (İncir Burnı). Simultaneously, Ottoman cartographers enriched their maps with local knowledge gathered through diplomatic and commercial networks, as demonstrated by their use of Slavic toponyms like "Dubrovenedik" for Dubrovnik, alongside the more common Romance (Venetian and Dalmatian) name "Ragusa." This dual

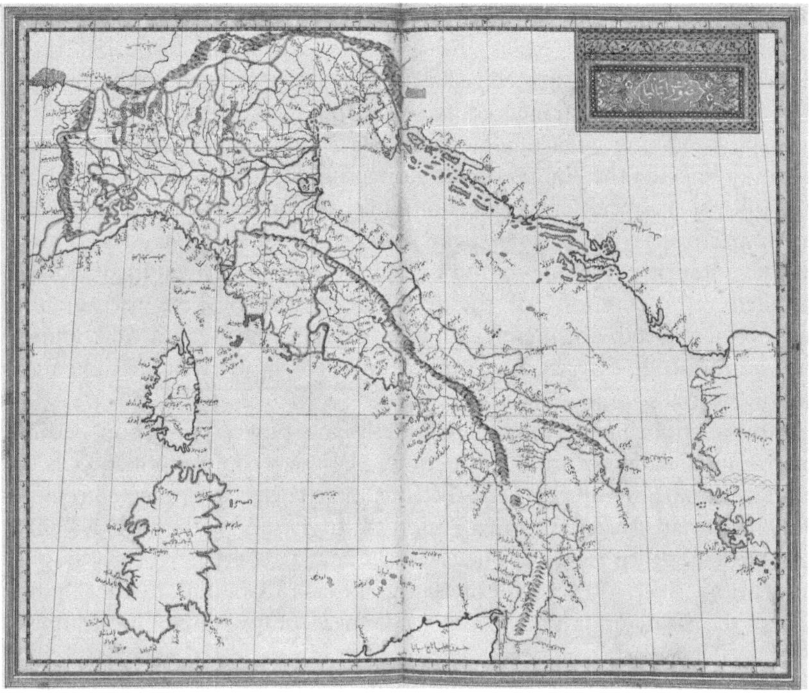

Fig. 8.11 Ebûbekir Behrâm b. ed-Dımeşkî, "Sûret-i İtalya," *Nusretü'l-İslam ve's-Sürûr fi Tahrîr-i Atlas Mayor* (Source: Istanbul University Central Library Rare Works, Turkish Manuscripts, 6606, folia 212b-213a)

approach reached its mature expression in seventeenth-century works like Kâtib Çelebi's *Müntehab-ı Bahriye*, *Cihannümâ*, and *Levâmi'u'n-Nûr* where Western cartographic precision merged seamlessly with Ottoman geographical understanding, creating charts that served both practical navigation and imperial administration.

8 Conclusions

The evolution of Ottoman cartography, particularly in its representation of the Adriatic Sea, reflects a cultural synthesis and practical adaptation. In its early stages, the Ottoman Empire predominantly relied on written geographical descriptions, adhering to the established scholarly traditions of the Islamic world. This approach, while thorough in its textual documentation, gradually evolved as the empire's expanding territorial reach demanded more sophisticated visual representations of geographical space. The sixteenth century emerged as a watershed moment in Ottoman cartography, marking what scholars consider a golden age in the field. This period witnessed the emergence of pioneering figures like Piri Reis, whose exceptional nautical charts revolutionized maritime cartography. His work, particularly the *Kitâb-ı Bahriye*, represented a crucial advancement in the systematic documentation of Mediterranean geography, including detailed representations of the Adriatic coastline. These cartographic achievements were not merely academic exercises but served vital military and administrative functions, supporting the empire's naval campaigns and territorial administration.

The relationship between cartography and power projection became increasingly evident during this period. Maps served as essential tools for military campaigns and territorial claims, particularly in contested regions like the Adriatic, where Ottoman interests intersected with those of Venice and the Habsburg Empire. The establishment of specialized cartographic institutions, such as the guild of cartographers in Istanbul, further underscores the strategic importance placed on accurate geographical knowledge. A particularly significant development was the emergence of scholars like Kâtib Çelebi, who played a pivotal role in bridging the gap between Eastern and Western cartographic traditions. Their work represented a conscious effort to synthesize diverse geographical knowledge systems, combining the best elements of both Islamic and European cartographic traditions. This synthesis was evident in works like *Cihannümâ* and *Levâmi'u'n-Nûr*, which marked the beginning of modern geographical understanding in the Ottoman Empire.

The Ottoman Empire's openness to learning from other cultures, particularly evident in its willingness to incorporate European cartographic innovations while maintaining its own traditions, proved crucial for maintaining its competitive edge in maritime affairs. This adaptability was reflected in the evolution of their nautical charts of the Adriatic Sea, which progressively incorporated more sophisticated technical elements while retaining distinctive Ottoman characteristics. This history of Ottoman cartography thus reveals not only the technical evolution of mapmaking but also the broader cultural and intellectual dynamics of an empire at the crossroads of civilizations. The development of their nautical charts of the Adriatic Sea, from initial written descriptions to sophisticated nautical charts, exemplifies the Ottoman Empire's remarkable ability to synthesize different traditions while serving practical imperial needs. This legacy continues to provide valuable insights into the relationship between cartography, power, and cultural exchange in the early modern Mediterranean world.

Acknowledgments The authors sincerely thank Josip Faričić for his careful review and invaluable contributions, particularly in identifying and verifying toponyms, linking Ottoman cartographic names to their modern equivalents, and providing insights into Venetian nautical charts. The authors are also grateful to Tome Marelić for his editorial contributions and to Dubravka Mlinarić for bringing this book to our attention and extending the invitation.

Primary Sources

Ahmed the Tangier (1413) Mediterranean Portolan (Manuscript No. 1823). Topkapı Palace Museum Library, Hazine Collection (TSMK, H.)

Al-Hac Ebu'l-Hasan (ca. 1560) Portolan Atlas (Manuscript No. 1822). Topkapı Palace Museum Library, Hazine Collection (TSMK, H.)

Ali Macar Reis. (16th century). Ali Macar Reis Atlası (Manuscript No. 644). Topkapı Palace Museum Library, Hazine Collection (TSMK, H.)

Atlas-ı Hümâyûn (Manuscript No. 1621). (seixteenth century). Istanbul Archaeology Museum Library

Blaeu J (1662) Atlas Maior. Labore and Sumptibus, Amsterdam

Cresques A (1375) Catalan Atlas [Map]. Bibliothèque nationale de France, Département des manuscrits (Espagnol 30)

Dulcert A (1339) Chart (C. Pl. Ge B 696 Res). Bibliothèque nationale de France

Ebûbekir Behrâm b. ed-Dımeşkî (1675–1685) Nusretü'l-İslam ve's-Sürûr fî Tahrîr-i Atlas Mayor (Manuscript No. 6606). Istanbul University Central Library Rare Works

Ibrahim al-Mursi (1461) Nautical Chart (Manuscript No. 882). Istanbul Naval Museum

Kâtib Çelebi (1645–1646) Müntehab-ı Bahriye (Manuscript No. 227). Presidency of Manuscript Institution of Turkey, Süleymaniye Manuscript Library, Aşir Efendi Collection

Kâtib Çelebi (1657) Tuhfetü'l-Kibâr fi Esfâri'l-Bihâr (Manuscript No. 1192). Topkapı Palace Museum Library, Revan Collection

Mercator G, Hondius J (1621) Atlas Minor Gerardi Mercatoris A. S. Hondio plurimis aenis Tabulis acutus et illustratus. Arnheim

Piri Reis (1521) Kitâb-ı Bahriye (Manuscript No. 337). Topkapı Palace Museum Library, Bağdat Collection

Piri Reis (1521) Kitâb-ı Bahriye (Manuscript No. 1). Presidency of Manuscript Institution of Turkey, Ali Emiri Manuscript Library, Coğrafya Collection

Ptolemy (1460s) Kitâbü'l-Coğrâfya fi'l-Ma'mûre mine'l-Arz (Manuscript No. 2610). Presidency of Manuscript Institution of Turkey, Süleymaniye Manuscript Library, Ayasofya Collection

Ptolemy (1460s) Kitâbü'l-Coğrâfya fi'l-Ma'mûre mine'l-Arz (Manuscript No. 2596). Presidency of Manuscript Institution of Turkey, Süleymaniye Manuscript Library, Ayasofya Collection

Unknown (n.d.) The Muhit Atlas. Biblioteca Nazionale di Napoli, BNN Ms. III.F.35, fol. 99

Walters Nautical Atlas (Manuscript No. W660). (1560's). Walters Collection

References

Al-Khaffaf E (2023) George Amirutzes's Arabic translation of Ptolemy's Geography made in 1465 at the order of Sultan Mehmed the Conqueror: The cultural aspect of the translation. In: Kaya MC, Özdemir N, Aksoy G (eds) The 2nd International Prof. Dr. Fuat Sezgin Symposium on History of Science in Islam Proceedings Book, pp 351–360

Aydın C, Aydın G (1992) Batlamyus. In TDV İslâm Ansiklopedisi, vol 5, İstanbul, pp 196–199

Barthe P (2008) An uncommon map for a common world: Hajji Ahmed's cordiform map of 1559. L'Esprit Créateur 48(1):32–44

Bekar C (2011) A New Perception of Rome, Byzantium, and Constantinople in Hezarfen Huseyin's Universal History (Unpublished master's thesis). Boğaziçi University, Institute of Social Sciences, Department of History, İstanbul

Benabbes M (2020) An Ottoman translation of Ptolemy's Geography made in the fifteenth century: The importance of an unknown work. In: Başar F, Kaçar M, Kaya C, Furat AZ (eds) The 1st International Prof. Dr. Fuat Sezgin Symposium on History of Science in Islam Proceedings Book, pp 417–427

Berardi L (2017) The sixteenth-century "Muhit Atlası": from a Venetian Globe to an Ottoman Atlas? Imago Mundi 69(1):37–51. https://doi.org/10.1080/03085694.2017.1242839

Brentjes S (2012) On two manuscripts by Abu Bakr b. Bahram al-Dimashqi (d. 1102/1691) related to W. and J. Blaeu's Atlas Maior. J Ottoman Stud 40:171–220

Brotton J (1997) Trading territories: mapping the early modern world. Cornell University Press, Ithaca, NY

Casale G (2012) From Hungary to Southeast Asia: The Ali Macar Reis Atlas in a global context. Osmanlı Araştırmaları 39(39):54–62

Evliya Çelebi (2006) Evliya Çelebi Seyahatnamesi: Topkapı Sarayı Kütüphanesi Bağdat 308 Numaralı Yazmanın Transkripsiyonu-dizini (vol 1, R. Dankoff, S. A. Kahraman, & Y. Dağlı, Eds.). İstanbul

Faričić J, Kljajić I, Mirošević L, Mlinarić D (2023) Symbolism of compass roses on early modern nautical charts of the Adriatic Sea. KN – J Cartography Geographic Inf 73:19–37. https://doi.org/10.1007/s42489-022-00129-z

Karamustafa AT (1992) Introduction to Ottoman cartography. In: The history of cartography, vol II. University of Chicago Press, Chicago, IL, pp 206–208

Karaçay Türkal M (2014) Hezarfen Hüseyin Efendi'nin Tenkîhü't-Tevârîhü'l-Müluk adlı eserine göre Tul u Arz ile Küre-i Arzın Inkisamı. In Uluslararası Katılımlı Osmanlı Bilim ve Düşünce Tarihi Sempozyumu Bildiriler Kitabı. 321–330

Kâtib Çelebi (2017) Levâmiu'n-Nur Fi Zulmet-i Atlas Minor: İnceleme-Tıpkıbasım (A. Üstüner & H. A. Arslantürk, Prep. by). Türkiye Bilimler Akademisi, Ankara

Kâtib Çelebi (2018) Tuhfetü'l-Kibâr fî Esfâri'l-Bihâr: İnceleme-Tıpkıbasım (İ. Bostan, Prep. by). Türkiye Bilimler Akademisi, Ankara

Mlinarić D (2011) The Novak Collection as an inspiration: characteristics of some rare maps of the northern and central Dalmatian coastline and the role of local cartographers. Kartografija i Geoinformacije 10(16):72–91

Mlinarić D, Faričić J (2024) The Eastern Adriatic Coast in the narratives of the early-modern Ottoman navigator and cartographer, Piri Reis. In: Ak M, Üstüner A (eds) Geography and Cartography in the Ottoman Empire. Istanbul University Press, Istanbul

Özdemir K (2007) Hezarfen Hüseyin Efendi'nin Tenkîhüt-Tevârîh adlı eserinin Selçukluların Zuhurundan Osmanlı Devleti'nin Kuruluşuna Kadar Geçen Bölümlerinin Transkripsiyon ve Değerlendirmesi (Unpublished master's thesis). Celal Bayar University, Institute of Social Sciences, Department of History, Medieval History Program, Manisa

Nas A (2019) Hezârfen Hüseyin Efendi'nin "Tenkîhü't Tevârih-i Mülûk" İsimli Eserinin Tahlili ve Metin Tenkidi (Unpublished master's thesis). Erzincan Binali Yıldırım University, Institute of Social Sciences

Sezgin F (1987) The contribution of the Arabic-Islamic geographers to the formation of the world map. Institut für Geschichte der Arabisch-Islamischen Wissenschaften an der Johann Wolfgang Goethe-Universität, Frankfurt am Main

Tunuslu Hacı Ahmed (2019) Dünya Haritası ve Coğrafya Risalesi (N. Karakaş Özür, Prep. by). Çizgi Kitabevi

Üstüner A (2017) Levâmi'u'n-Nûr (Metin-Değerlendirme) (Unpublished doctoral dissertation). Istanbul University, Institute of Social Sciences, Department of Turkology Studies, Istanbul

Üstüner A (2019) Klasik Çağ sonrası Osmanlı coğrafya çalışmaları. Türkiye Araştırmaları Literatür Dergisi 17(33):41–104

Üstüner A, Çinar M (2024) Fatih Sultan Mehmed Devri Haritaları ve Ptolemaios. In Din ve Coğrafya Sempozyumu II: 4/10 Asır Sonrası İslam Coğrafyacılığının Tarihsel Gelişimi Bildiri Kitabı. Çorum: Hitit Üniversitesi Yayınları. 447–470

CHAPTER 9

Cartometric Analyses: Methodological Issues and Their Significance for the History of Cartography and Historical Studies in General

Tome Marelić

1 The Outline

The cartometric approach itself is a set of methods and tools used for processing quantitative historical data that are either integral components of old maps or charts or exist in some other related data formats. Yet, its proper utilisation requires possessing an additional body of knowledge and skills regarding both sciences and humanities, such as mathematics, geodesy, the history of science, the history of navigation, etc. Otherwise, there is a likelihood that cartometric analyses will be conducted in a methodologically inaccurate manner, that conclusions derived from those computations will be erroneous and scientifically invalid, or both.

This volume permits going into all of the subsidiary disciplines and techniques. Consequently, its content is confined to providing essential

T. Marelić (✉)
Department of Geography, University of Zadar, Zadar, Croatia
e-mail: tmarelic@unizd.hr

J. Faričić, T. Marelić (eds.), *Early Modern Nautical Charts of the Adriatic Sea*, https://doi.org/10.1007/978-981-96-8046-7_9

information related to the existing cartometric analyses, some important methodological issues, and how they may contribute to a better understanding of old maps and charts, the development of the history of cartography as a discipline, and broader insights into human history. The chapter is crafted to be accessible to a broad array of readers, regardless of whether they are coming from history of cartography background or not. In order to make the concept and techniques of cartometric analyses more approachable to newly interested readers, certain elements are deliberately mentioned multiple times across different paragraphs and explained in a more simplistic manner. Figures derived from the author's previous cartometric studies supplement the text by vividly and intuitively illustrating its key elements. Another volume-induced limitation relates to production periods and coverage of old maps and charts. The cartometric methods described in this chapter pertain to nautical charts and geographical maps created during the late medieval and early modern periods, before the conduction of geodetic and hydrographic surveys, and the examples provided cover the Mediterranean area and the Adriatic Sea, depending on the topic of certain passages. This is because the impact of choosing a certain map projection as a frame of reference is better understood in the example of the entire Mediterranean, whereas some approaches for processing spherical coordinates are better demonstrated by using the smaller area as an example. Since the author's research focuses primarily on the geometry of portolan charts, the majority of its examples and illustrations naturally belong to that specific sub-niche within the history of cartography.

The chapter does not contain one section solely dedicated to previous research. It was intentionally written in such a way that certain elements of the existing cartometric studies conducted by the author and other scholars are dispersed across the respective (sub-)topics. These are partially explained and addressed via cross-references, enabling the reader to search the missing information externally. The final section emphasises the applicability of cartometric analysis results, which, when coupled with other historical documentation, may operate as forensic tools that strengthen understanding of both the history of cartography and human history overall. Since cartometric analyses are not applicable to every study within the history of cartography because not all old maps were made with a realistic display of the (known) world in mind, the first logical step is to provide the reader with a brief outline of cartographers' approach to planimetric accuracy in mapmaking across certain important historical periods.

2 A Brief History of Planimetric Accuracy in Mapmaking

Between classical antiquity and the advent of modern geodetic and hydrographic surveys that took place in the eighteenth and nineteenth centuries, the attention to planimetric accuracy of old maps and charts went through numerous changes. The mathematical approach to mapmaking that included estimating the size of the Earth, defining coordinates of locations (their latitudes and longitudes) within the spherical geometry, and the development of map projections was thoroughly practised during classical antiquity.[1, 2] In contrast, the *mappae mundi* that were made during the early Middle Ages in Western Europe to depict the Christian understanding of the known world were more elementary in design (Woodward 1987), whereas in the ninth and tenth centuries, the Islamic cartography adopted a more scientific approach, inspired by classical antiquity sources that were, at the time, freshly discovered in the former Byzantine provinces (Tibbets 1998).

The most enigmatic and, perhaps, one of the most important periods in the history of cartography occurred in the late thirteenth century, when an entirely new branch of cartography, nautical cartography, was established, as evidenced by the sudden appearance of portolan charts in the Mediterranean. It is due to the extremely realistic appearance of their coastlines and a complete historical lack of their evolutionary path because the oldest known samples have already been made to a fully developed stage (Campbell 1987; Edson 2007; Nicolai 2014, 2024; Marelić 2024a, b).[3]

The rediscovery of Claudius Ptolemy's book *Geographike Hyphegesis* (commonly known as *Geography*, originally written in the second century

[1] The most renowned calculations of the Earth's great circle in classical antiquity were performed by Eratosthenes in the third century BCE and by Posidonius of Rhodes in the second and first centuries BCE (Russo 2004).

[2] Marinus of Tyre from the first century CE is the presumed inventor of the equidistant cylindrical projection, whereas in the second century CE Claudius Ptolemy described how equidistant conic and pseudoconic projections should be constructed (Keuning 1955; Snyder 1993).

[3] The earliest-known portolan charts include the anonymous *Pisane* (see Figs. 9.1 and 9.2), *Cortona*, *Avignon*, and *Lucca* charts made in the late thirteenth or early fourteenth century, and those made by Pietro Vesconte (see Fig. 9.3) between 1311 and 1327 (Nordenskiöld 1897; Campbell 1987; Billion 2011; Mille 2023). The anonymous *Carta Riccardiana* (c. 1300–1325) was most likely drawn by Vesconte as well (Marelić 2024b).

CE) with a list of around 8000 locations, 6300 of which were designated by their latitudes and longitudes, had the greatest impact on European cartography in the early Renaissance (Berggren and Jones 2000; Gautier-Dalche 2007). The use of spherical coordinates to determine point locations—the so-called geometrisation of space, according to Richard L. Kagan and Benjamin Schmidt (2007: 663)—and the implementation of graticules on maps that were deliberately made in the map projection was a novel approach for Western European cartographers at the time, who initially followed those principles to create their geographical maps. During the early modern period, prior to the conduction of extensive geodetic and hydrographic surveys, the nautical charts that were produced in parallel with geographical maps and geometry-wise noticeably differed from them (Marelić 2023a, b) had also become supplemented with those cartographic elements, although their authors appear to have been more conservative in terms of their implementation, adopting them more cautiously at a somewhat slower rate.

3 The Essentials of Cartometric Analysis

Simpler types of old maps, such as *mappae mundi* and their variant, the T-O maps, can only be analysed descriptively because their creators did not prioritise the metrically accurate representation of spatial relationships, whereas some more intricate works, such as the so-called *Tabula Peutingeriana*,[4] allow only a limited quantitative analysis. This is because the distance annotations accompanying its complex road network are the only exact quantitative historical data that seems to have been observed in the field, unlike the overall image of the area and the road network itself, which are longitudinally extremely compressed to fit within its 6.7 × 0.3-meter format, designed to be rolled up and carried around as a scroll when not in use.

On the other hand, the extremely accurate display of coastlines on portolan charts has prompted a scholarly approach to them from the late nineteenth century. Matteo Fiorini, for example, assumed that they were made in the equidistant azimuthal projection because the colour-coded lines of their wind roses radiate from a central point (Fiorini 1881), but

[4] The anonymous map portraying a graphic representation of a Roman *itinerarium* (travel guide), dated to the early thirteenth century, whose original was likely made in the fourth or fifth century CE (Dilke 1985).

Hermann Wagner's first-ever cartometric analysis in 1896 explicitly refuted this assumption. Wagner coined and proposed the term *cartometric method*, examined portolan charts created by Pietro Vesconte (1318) and Gratiosus Benincasa (1480), and concluded that their metrics are very similar to a normal cylindrical projection and that they are most likely mosaic images composed of sub-parts that were graphically stitched together (Wagner 1896/1969). Wagner's methods were analogue and rudimentary in comparison to modern standards, but his revolutionary conduct exposed a completely new and previously unknown realm of methodological approaches to be taken within the history of cartography niche. Despite advancements in computer sciences and engineering since Wagner's study that lifted cartometric analyses to a whole new technological level, and some scholars pointing out that cartometric methods are beneficial to avoid generating a "heavy load of theory ... on a visual impression" (Skelton 1965: 6), the number of cartometric studies of old maps is still much smaller compared to those that emerged from the traditional descriptive approach. Some notable and representative cartometric studies will be mentioned and described in the following paragraphs, based on their respective sub-topics.

Within the context of the history of cartography, cartometric methods are essential tools for extracting geometric and other quantitative data from historical sources that are often invisible to the naked eye and comparing them directly to other similar datasets (whether historical or modern), and are usually conducted with the application of GIS and statistical software. Although the input data in such analyses are primarily cartographic, as they are typically encapsulated within frameworks of old maps and charts, some other relevant quantitative input data formats, such as coordinates, distances, or bearings contained in textual sources, may be imported in GIS software and cartometrically analysed as well. The basic concept is to place historical input data—whether in the form of digitised reproductions (images) of old maps and charts or spreadsheet data containing other significant metrics—in direct geometric relationship with their reference modern equivalents, and the entire process consists of multiple consecutive steps. The initial step involves acquiring and creating the necessary reference data (treated as error-free) and properly identifying the required historical data (referred to as the *identical data*) to pair with them. In the case of old maps and charts, those are identical points typically utilised for the subsequent step: their georeferencing.

Georeferencing (Fig. 9.1) is a process conducted in GIS software within which a sufficiently redundant number of identical points identified on the old map or chart are paired with their reference counterparts on a modern map in a selected map projection and geometrically transformed. The geometric transformation enables the digital image of the old map and its identical points to be shifted, scaled, and rotated to obtain their geometric best fit to the reference map projection by achieving the *least-squares estimation* (LSE) of their axial *residuals* (*dX*, *dY* in kilometres). These are the displacements of identical points that occur along its X and Y axes when the geometry of an old map differs from the geometry of a reference map (Fig. 9.2). In addition, the residuals need to be de-projected, that is, translated into their spherical or ellipsoidal angular displacements ($d\lambda$, $d\varphi$ in degrees). The following step is to convert them into distances along the longitudinal and latitudinal arcs between the reference–identical point pairs (*dLON*, *dLAT* in kilometres). The final step is the computation of axial de-projected *root mean square errors* of residuals (RMSE *dLON*,

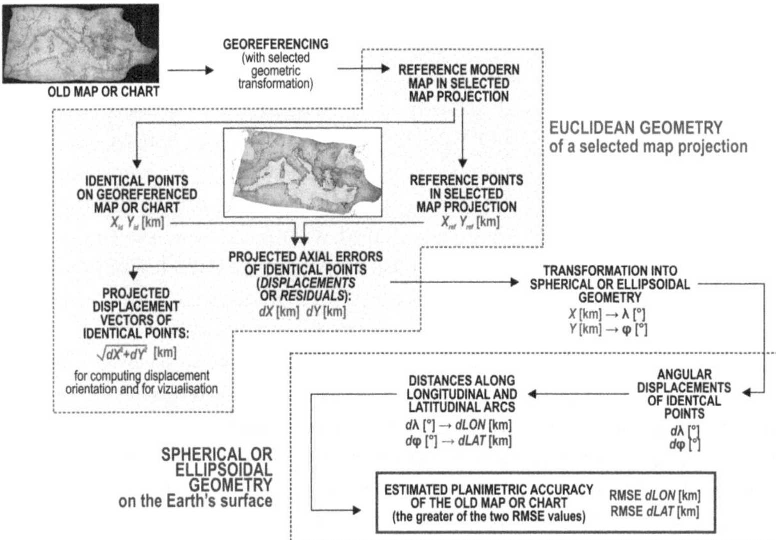

Fig. 9.1 The simplified schematic of the georeferencing process and the accuracy estimation of old maps and charts by cartometric means (Portolan chart source: *Bibliothèque nationale de France*, GE B-1118 (RES))

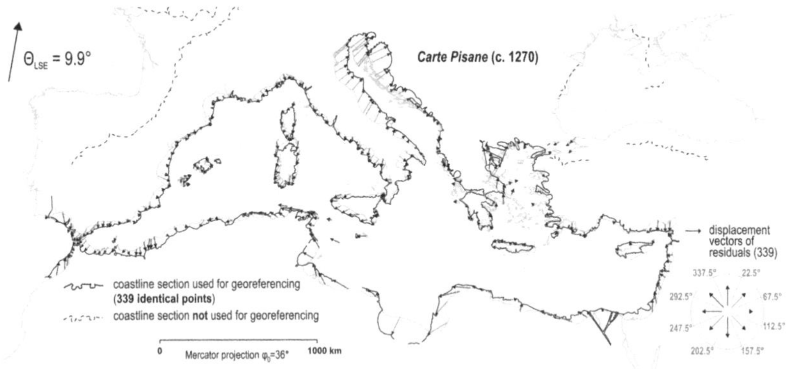

Fig. 9.2 The magnitude and orientation of displacement vectors of residuals of the anonymous *Carte Pisane* (c. 1270, the oldest known portolan chart) georeferenced to a modern map in the Mercator projection across the Mediterranean area. Basemap shapefile source: marineregions.org (Claus et al. 2017)

RMSE *dLAT* in kilometres) (Fig. 9.1). The greater of the two values (whether RMSE *dLON* or RMSE *dLAT*) is used to indicate the estimated planimetric accuracy of the entire map (Jenny and Hurni 2011; Nicolai 2014; Penzkofer 2016). Since these, in essence, represent the mapping errors, the accuracy of the map is understood to be greater if the RMSE values are smaller and vice versa. Also, when a single map is georeferenced to a modern map iteratively, each time using a different map projection (see Fig. 9.3), smaller RMSE values indicate a greater likelihood of the map being originally made in such a projection.

4 Some Important Methodological Issues

Although the georeferencing of the old map or chart seems relatively straightforward on the conceptual level, there are several important parameters that need to be considered during the procedure, since improper configuration might negatively affect both the outcome and the interpretation of the results. The first two issues are the parameters of the coordinate reference system and the geometric transformation to be used because they strongly affect the results in the early stages of the research. The third issue includes transforming projected values back to spherical or ellipsoidal geometry in order to eliminate the distortions induced by the systemic properties of the analytical framework itself.

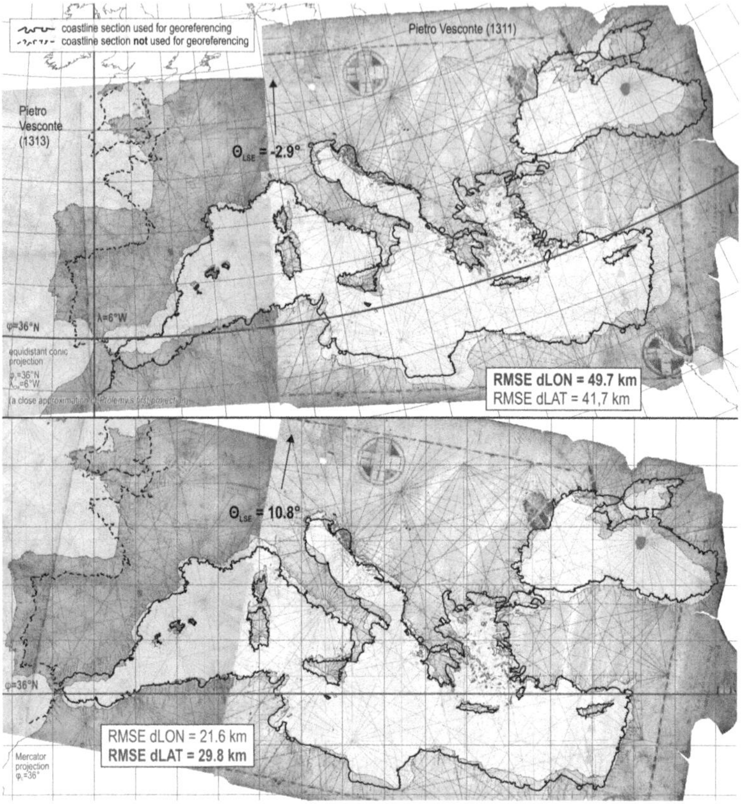

Fig. 9.3 The manually assembled composite of two Pietro Vesconte's portolan charts georeferenced to the modern map in the close approximation of the *Ptolemy I* projection centred near Gibraltar (the upper part), and to the modern map in the Mercator projection (the lower part), with their vectorised coastlines and de-projected accuracy values. The greater of the two axial values in bold indicate the estimated planimetric accuracy of the entire georeferenced unit (Portolan chart sources: *Archivio di Stato di Firenze*, CN 01; *Bibliothèque nationale de France*, CPL GE DD-687 (RES)). Basemap shapefile source: marineregions.org (Claus et al. 2017)

Those issues are important because the georeferencing process operates by comparing the image of the identical dataset to the image of the reference dataset within the same Euclidean environment, in accordance with which the geometry and visual appearances of both the reference and identical images are defined. However, the coordinates on the Earth's surface, as well as bearings and distances between those points observed during surveys, are established according to either spherical or ellipsoidal geometry. This means that the geometry of maps and charts that were deliberately made in a map projection is determined with its geometric parameters, that certain geometric features from the real world (whether distances, shapes, or areas) become distorted by definition, and that their distribution and magnitude vary across the map field as well.

4.1 *Coordinate Reference System (CRS)*

The first important methodological issue is that—unlike the real world—the geometry of all the data within the GIS framework is typically Euclidean and determined by the user-defined projected coordinate reference system (CRS). In other words, the structure of the whole dataset, including its visual appearance and its geometric properties, is defined in accordance with the selected map projection and geodetic datum. This means that the coordinates of points on the Earth's surface (λ, φ in degrees) are, by definition, transformed into Euclidean coordinates (X, Y in kilometres) depending on the selected projected CRS, and that the geometry of all the identical datasets imported into that framework will likewise be expressed in accordance with the Euclidean parameters of the selected CRS. In many types of GIS software, the parameters of the CRS can be altered any time, meaning that selecting a new CRS or modifying the existing one will geometrically alter the entire reference dataset, and that a completely new set of Euclidean coordinates will be assigned to it.[5]

The user has to be well aware of the geometric properties of the CRS framework in advance because wrongly set parameters can easily lead to incorrect results, especially if the research area stretches across large territorial extents, making it more prone to projection-induced distortions. For example, if one is interested in computing certain area-values by using GIS software, the selected projected CRS should be based on some equal-area projection,[6] whereas in cases of studying bearings or distances along the rhumb lines (loxodromes), one should opt for a CRS that is based on

[5] The upper part of Fig. 9.3 shows the modified map projection within the CRS: the equidistant conic projection was modified to become centred at the longitude near Gibraltar (λ = 6°W), and φ = 36°N was set as its single standard (true-to-scale) parallel.

[6] Some typical examples of equal-area map projections are the Mollweide pseudocylindrical projection, the Albers conic equal-area projection, and the Lambert azimuthal equal-area projection (Snyder and Voxland 1989).

the Mercator projection.[7] Otherwise, the computed geometric parameters become erroneous, and conclusions derived from them are incorrect and invalid. This is especially important for the users of *MapAnalyst*, which is a relatively simple and user-friendly software for elementary cartometric analyses of old maps (Jenny 2006, 2010; Jenny et al. 2007; Jenny and Hurni 2011). The software automatically links identical points on the old maps or charts to *OpenStreetMap*'s reference map in the Mercator projection. Those points are, then, de-projected and projected again, but this time in the transverse cylindrical equal-area projection centred at the mean longitude of all control points. This choice of a reference map is suitable only for studying smaller areas because the distortions of distances and bearings become more and more extreme on the peripheral areas as the longitudinal extent of the research area increases (Nicolai 2018).

In cases where the old maps or charts are the result of some systematic field survey and their original geodetic parameters are well known, the CRS of the reference dataset should be set in the same way. If those parameters are unknown, and if the map projection of the old map or chart cannot be estimated plausibly enough with the naked eye (as is the case with portolan charts), the cartometric analysis necessarily becomes an iterative process in which the same old map or chart is georeferenced multiple times to the same reference dataset, but each time with a different CRS assigned to it (Fig. 9.3). This means that every iteration (georeferencing of the same map or chart to the same reference dataset using a different CRS in each session) yields different estimated planimetric accuracy values. Smaller RMSE values at the end indicate greater geometric similarities between the old map or chart and a certain map projection and vice versa, meaning that this iterative process can be used to determine the most plausible map projection of an old map or chart.

4.2 *Two-Dimensional Geometric Transformations*

There are several two-dimensional geometric transformations to be selected for georeferencing, ranging from simpler to more complex, each affecting the geometry and the image of the georeferenced map in its own way. In

[7] This is because, unlike the stereographic or Lambert conformal conic projection, the Mercator projection does not display the convergence of the meridians and therefore retains the same direction of north across the entire map field, and plots loxodromes as straight lines. If distances along the rhumb lines on the Earth's surface are required for the cartometric analysis, the prerequisite step is to de-project their (Euclidean) values defined within the Mercator CRS.

general, the more complex the transformation, the more adjustments the original map image undergoes during the georeferencing process.

For example, an eight-parameter projective transformation, which is based on a more complex computation and creates a perspective two-dimensional projection of a two-dimensional image (resembling its visual appearance when observed from a different angle), was used by Scott A. Loomer in the initial stage of his cartometric study of portolan charts (Loomer 1987). He applied it to the (invisible) circles of the wind roses on portolan charts—used as proxies—to convert them into perfect circles in order to rectify the perspective viewshed of cameras used to photograph them and to reduce the eventual distortions of their carrier material, the parchment, that might have occurred over centuries.

After Loomer made those corrections to prepare portolan charts for the analysis, he applied a four-parameter Helmert similarity transformation to cartometrically examine 26 portolan charts in comparison to nine map projections. Helmert transformation uniformly scales and rotates both axes of the unit (Modenov and Parkhomenko 1965), preserving the geometry of the georeferenced image of an old map or chart in relative terms. It was also used by Tome Marelić in his cartometric studies of old nautical charts and geographical maps of the Adriatic Sea (Marelić 2022, 2023a, b) and portolan charts showing the Mediterranean and Black Sea (Marelić 2024a, b). A tangible example can be seen in Fig. 9.3 that shows how the manually assembled composite of Pietro Vesconte's two portolan charts, made in 1311 and 1313 (the former of which is the oldest known portolan chart whose author and year of production are known explicitly),[8] was iteratively georeferenced to the same reference dataset twice, by using different CRS each time. Although the geometry of the reference dataset significantly differs in each iteration, leading to different scales and rotations (θ_{LSE}) of the georeferenced images, their geometry remained relatively unaltered in both cases because the same four-parameter similarity transformation was applied each time.

In his doctoral thesis, Roel Nicolai analysed portolan charts using a six-parameter affine transformation (Nicolai 2014) by which each axis becomes scaled and rotated independently. The same transformation was used by Evangelos Livieratos and Chrysoula Boutoura in their cartometric

[8] On its eastern margin it reads: *Petrus Vesconte de Janua fecit ista carta ann[o] dni MCCCXI*, meaning *Pietro Vesconte of Genoa made this chart in 1311* (Nordenskiöld 1897: 57).

study of the *Carte Pisane* (Livieratos and Boutoura 2018). In his recent study, Nicolai used a five-parameter affine transformation by which each axis is scaled independently, whereas both of them are uniformly rotated (Nicolai 2024). Affine transformations allow more degrees of freedom in comparison to a Helmert transformation, meaning that they deform the image of georeferenced units if there are significant geometric differences between them and the reference map, and yield greater estimated accuracy values at the end.[9] They are exceptionally useful when the geodetic parameters of old maps and charts are already known and are intended to be georeferenced to a reference dataset (CRS) of the same geometric properties.

4.3 The Necessity of De-Projecting the Datasets

Since georeferencing is performed within the constraints of the Euclidean geometry, the accuracy values derived from it are not directly applicable to the real world because each map projection (defined within the selected CRS) distorts certain parameters of spherical or ellipsoidal geometry of the Earth in its own way across the map field. For example, a portion cut from a sphere along two meridians and stretching between poles is a digon, a polygon made of two straight lines on the spherical surface that meet at two points (vertices) at a certain angle, and which cannot exist in the Euclidean geometry because its most basic polygon is a triangle. Moreover, each of the three angles in the equilateral triangle in the Euclidean geometry has 60°, whereas in the spherical geometry the equilateral triangle has internal angles of 90°.

Because accurately made maps and charts are, in essence, the projected imagery of observations that were initially made on the Earth's surface, the aftermath of the cartometric analysis should be directly applicable to

[9] The number of degrees of freedom of different two-dimensional transformations affects the computation of their RMSE of residuals. In the case of a four-parameter Helmert transformation, the formulae are: $RMSE\ dLON = \sqrt{\frac{\sum_{i=1}^{n}(dLON_i)^2}{2n-4}}$ and $RMSE\ dLAT = \sqrt{\frac{\sum_{i=1}^{n}(dLAT_i)^2}{2n-4}}$, whereas in a six-parameter affine transformation those are: $RMSE\ dLON = \sqrt{\frac{\sum_{i=1}^{n}(dLON_i)^2}{2n-6}}$ and $RMSE\ dLAT = \sqrt{\frac{\sum_{i=1}^{n}(dLAT_i)^2}{2n-6}}$.

the Earth's surface as well. This means that after the georeferencing is done, the results need to be returned into spherical or ellipsoidal values using the inverse process in order to emulate how the planimetric accuracy of maps and charts translates into displacements of identical points in the real world. Therefore, their Euclidean residuals (*dX*, *dY* in kilometres) need to be de-projected by all means, and the intermediate step for doing so is converting them into angular differences between identical-reference point pairs (*dλ*, *dφ* in degrees).[10] The final step is to recompute those angular differences into axial distances along the longitudinal and latitudinal arcs on the Earth's surface (*dLON*, *dLAT* in kilometres; see Fig. 9.3).[11] In 2023, Marelić published a study that illustrates the necessity of such a procedure, even in the case of a relatively small research areas. He cartometrically examined 11 early modern manuscript and printed nautical charts of the Adriatic Sea basin in comparison to the Mercator projection and the equidistant cylindrical projection, both set with $\varphi_0=36°$ as their standard, true-to-scale parallel (Marelić 2023a). The research reveals that nine examined charts—the majority of which do not contain graticules, and some of which are manuscript portolan charts—showed better fit to the Mercator projection, which is methodology-wise important for two main reasons. First, the true-to-scale parallel of both reference maps is located south of the Adriatic Sea (with latitudinal extent between $\varphi = 39.8°$N and $\varphi = 45.7°$N, and mid-latitude of around $\varphi = 42.7°$N), meaning that both projections induce distance distortions across the entire research area. Second, the same true-to-scale parallel of $\varphi_0 = 36°$ for both reference maps dictates that the equidistant cylindrical projection retains the same true-to-scale renderings of latitudes (distances along its images of meridians), whereas the Mercator projection progressively distorts distances towards the (north) pole. This means that all the cartometrically generated data will, in most cases, 'favour' the equidistant cylindrical projection unless they are de-projected.

[10] The majority of GIS software allow direct translation of projected (Euclidean) coordinates into spherical or ellipsoidal coordinates (depending on the CRS), and vice versa.

[11] If the CRS contains a WGS84 ellipsoid, the following formulae can be used: | *dLON* | [*km*] = | *d* λ | [°] × (π × 6378.137 × cos (*mid LAT*))/(180 √ (1 – 0.006 × *sin* ^ 2(*mid-LAT*))) and | *dLAT* | [*km*] = | *d* ϕ | [°] × (111.132 – 0.559 × cos (2 × *mid-LAT*) + 0.001 × cos (4 × *midLAT*). The *midLAT* [*rad*] represents the mid-latitude between the reference point and its corresponding (georeferenced) identical point.

5 Cartometric Analyses of Historical Coordinates

The aforementioned characteristics of the georeferencing process are purely geometric in essence, which means that either no coordinates were assigned to their identical points by the cartographers who created them, or that they were intentionally ignored during the analysis. It is, for example, the only method for cartometrically analysing portolan charts. However, many later-made nautical charts, and especially the geographical maps, have been deliberately made in the map projection and contain graticules of longitudes and latitudes. In such circumstances, the identical points are (historical) coordinates that were assigned to them by their authors for their own reasons. Furthermore, some historical records, such as Ptolemy's well-known *Geography* (being the sole such survivor of classical antiquity), contain coordinates in textual format. Cartometric analyses can also be applied to historical records containing coordinates (Fig. 9.4), albeit the methodology somewhat changes in comparison to the purely geometric approach.

The first important difference is that historical coordinates, whether an integral part of old maps or textual in nature, are not imported into GIS software as an image but as a point dataset. This means that the initial step is to (manually) extract them one by one and record them on properly designed spreadsheets within which every point receives its proprietary historical longitude and latitude. Those spreadsheets can then be imported into GIS software set with the geographic CRS (made only of Earth's approximation as a sphere or some reference ellipsoid). In such a case, spreadsheet values become plotted as point datasets whose every point is positioned (georeferenced) individually according to its previously designated coordinates, which represents the second major difference in comparison to the purely geometric approach. This is because, unlike georeferencing an entire image of a map or chart at once, whose end result is determined by the parameters of the selected transformation, each point is imported and georeferenced within the GIS framework on its own, and no two-dimensional geometric transformation is tethered to the process itself. In later stages, those point datasets are usually projected by assigning some map projection to the CRS and reprojected by changing the map projection if needed. By doing so, the spatial layout of points (the geometry of the entire point dataset) changes in accordance with the selected map projection.

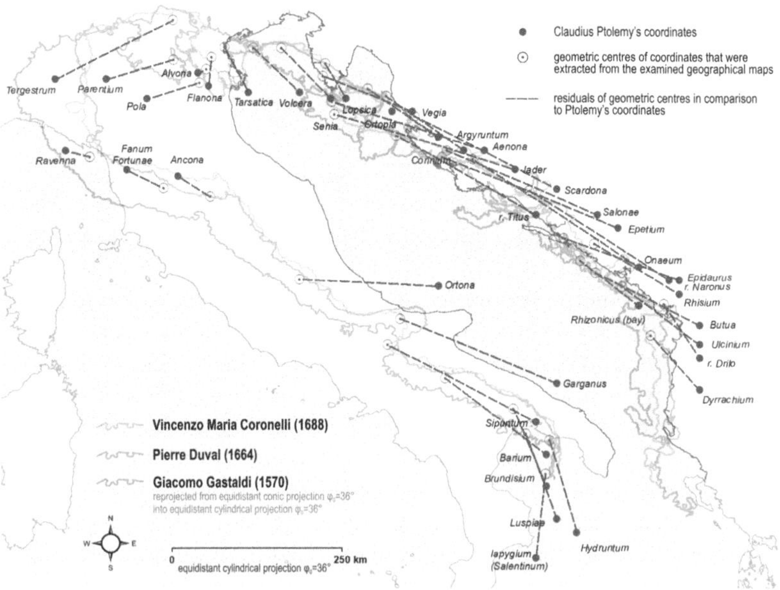

Fig. 9.4 The displacement vectors of residuals of the geometric centres of coordinates on three geographical maps (georeferenced by using the coordinates extracted from them) in comparison to Claudius Ptolemy's coordinates of locations on the Adriatic Sea coast. All datasets were longitudinally shifted by $\Delta\lambda_0 = -25.2°$. Ptolemy's coordinates source: Stevenson 1991. Basemap shapefile source: marineregions.org (Claus et al. 2017)

It is usually not possible to calculate the accuracy of historical coordinates in their current form because their longitudes have been expressed in accordance with another (historical) prime meridian. For example, Claudius Ptolemy's prime meridian passed through the so-called *Fortunate Islands* or the *Islands of the Blessed*, whose location is thought to be in the Canary archipelago. At the time, it was considered the westernmost limit of the known world (*oikoumene*), and Ptolemy chose it so that all the longitudes pointed in the same direction, ranging from 0° to 180°E (Berggren and Jones 2000). This means that historical longitudes should be translated into modern standard (recomputed with respect to the Greenwich Prime Meridian) as accurately as possible. If not, the entire historical point dataset is significantly shifted to the east or west, and the longitudinal accuracy cannot be calculated reliably. Because these point datasets represent coordinates believed to be true by the cartographers who made old maps and charts, they can be treated as pseudo-reference points and used to georeference the image of an old map or chart to them (Figs. 9.4 and 9.5).

Cartometric analyses of Ptolemy's coordinates were conducted by Livieratos and his associates (Livieratos et al. 2007, 2008), Christian Marx (2011, 2012, 2016), Irina Tupikova (2014), Marelić (2023b), and Jan Martínek and Aleš Létal (2023). In addition, Livieratos cartometrically compared the geometric properties of two *Ptolemaic maps* of Crete made in the fourteenth and seventeenth centuries[12] with Ptolemy's coordinates for the same area and concluded that there are significant geometric similarities between the historical datasets that were made to different formats (Livieratos 2006). Marelić's study (Marelić 2023b) shows that the coordinates of printed geographical maps of the Adriatic Sea and the surrounding areas made in the sixteenth and seventeenth centuries are largely similar to Ptolemy's coordinates dataset, retaining its renowned longitudinal exaggeration (Fig. 9.4). In contrast, the coordinates on the contemporaneously made nautical charts followed a completely different lineage, with longitudes that were seemingly deduced from the graphical appearance of coastlines on portolan charts (see Fig. 9.5).

6 Cartometric Analyses as the Forensics of Human History

The most obvious importance of cartometric analyses for the history of cartography as a discipline is that their application explicitly exposes metrics that are encapsulated within the historical cartographic sources but are not deducible with the naked eye, and from the phenomenological standpoint, modern cartometric analyses literally pose as the computer-aided forensics of the historical data. There is a plethora of their applications, many of which were already mentioned, and their utility covers both scientific and everyday purposes. One praiseworthy example of georeferencing old maps for everyday use is the *Arcanum project*, whose end result is a web portal that contains numerous old topographical, cadastral, and

[12] *Ptolemaic maps* are late medieval and early modern attempts to reconstruct the original cartographic content of Claudius Ptolemy's *Geographike Hyphegesis* which has not been preserved (Gautier Dalché 2007).

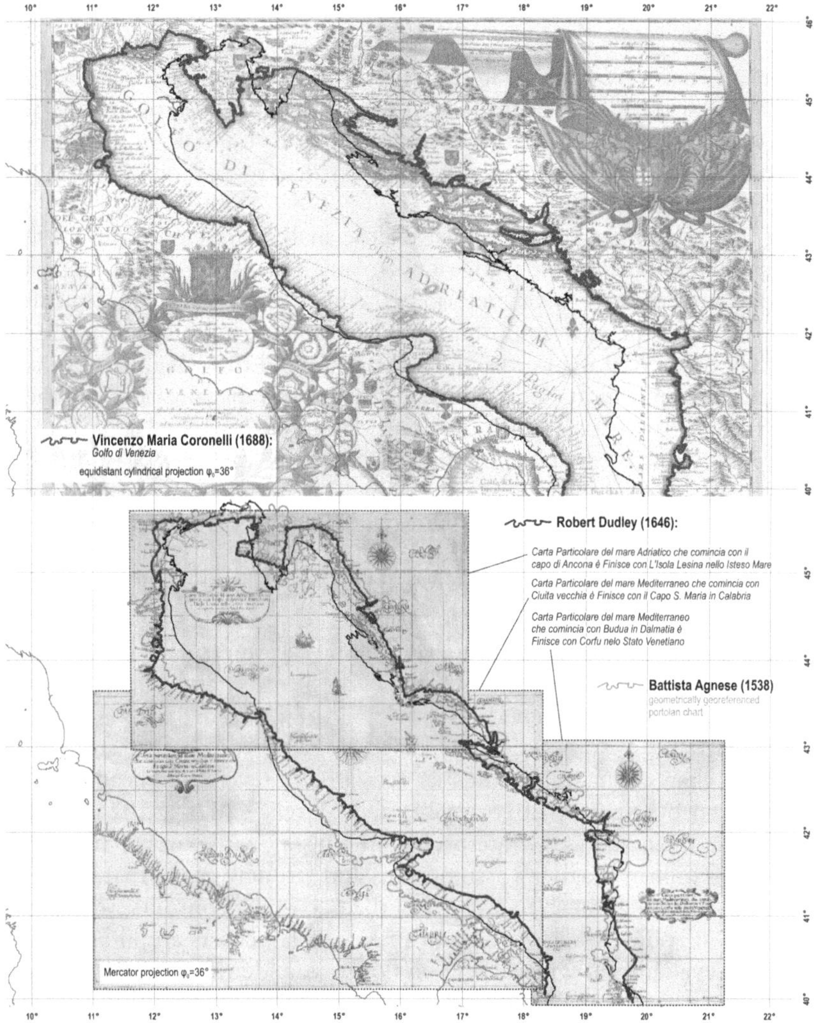

Fig. 9.5 Vincenzo Maria Coronelli's map from 1688 georeferenced to a modern map in the equidistant cylindrical projection φ_0 = 36° (the upper part), and the composite of Robert Dudley's three nautical charts from his 1646 book *Dell'arcano del mare* georeferenced to a modern map in the Mercator projection φ_0 = 36° (the lower part). Both units were georeferenced by using the coordinates extracted from them and with the application of proprietary longitudinal shift ($\Delta\lambda_0$) (Old map and nautical chart sources: *Nacionalna sveucilišna knjižnica u Zagrebu, zbirka karata i atlasa*, S-JZ-XVII-56; *National Library of Finland, Nordenskiöld collection* NBN:fi-fe201002051338). Basemap shapefile source: marineregions.org (Claus et al. 2017)

city maps that were scanned in high resolution and georeferenced within a web-GIS server.[13] The platform is accessible through desktop and mobile interfaces and allows the user to observe the historical data in their topologically correct order and to zoom in on desired sections for a more detailed view.

From a scholarly standpoint, cartometric analyses have proven to be the indispensable tools for the study of portolan charts. Wagner's 1896 paper not only introduced a novel method in the history of cartography, but also sparked one of the most important discoveries in human history by revealing their cylindrical map projection properties and the possibility of their sub-piece composition, and posting the hypothesis that the origin of their spatial data may be older than previously thought (Wagner 1896/1969). Regrettably, his peers, largely unfamiliar with his cartometric approach, failed to acknowledge Wagner's incredible feat (Winter 1948, 1956), and a significant segment of historians of cartography continues to express scepticism towards such concepts even today. Between Wagner's work and Loomer's extensive study that exposed their geometric similarity to the Mercator projection but did not investigate their possible origin in greater detail (Loomer 1987), there were only a few cartometric studies that are either simplistic or include only a small sample of charts (Clos-Arceduc 1956; Tobler 1966; Duken 1988).

The two most extensive and detailed ongoing cartometric studies of portolan charts, which have thus far appeared in several publications, are undertaken by Nicolai (2016, 2024) and Marelić (2024a, 2024b). The results of their analyses—interpreted in unison with known historical sources—confirm Wagner's assumptions and strongly argue that the spatial data used to establish the shapes of their coastline contours were likely acquired at some earlier period and that portolan charts are, most likely, late medieval and early modern copies of some older maps or charts. For example, the clockwise rotations of their georeferenced images in comparison to the Mercator projection, which occur due to anticlockwise-tilted renderings of their coastlines, do not agree with the regional distribution of the magnetic declination across that area at the time of their creation according to the CALS3k.4 paleomagnetic model.[14] This means that the cartometric evidence explicitly refutes the possibility that

[13] Georeferenced historical maps on the *Arcanum* webpage can be accessed via the following link: https://maps.arcanum.com.

[14] For more information about the CALS3k.4 model, see Korte and Constable (2011).

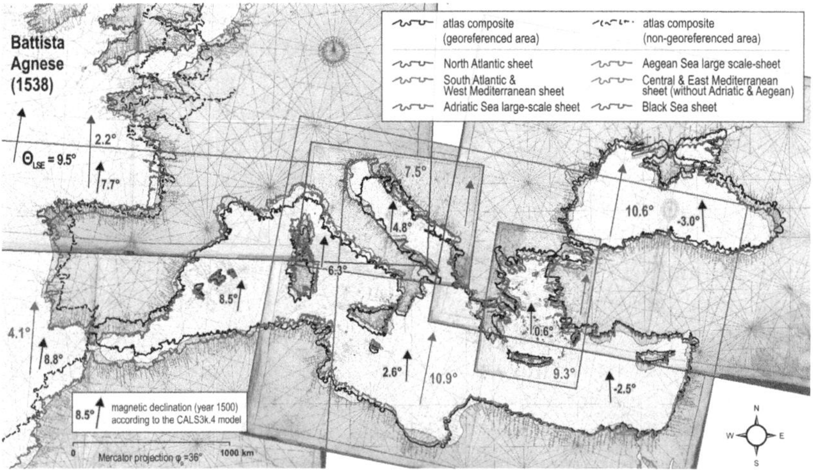

Fig. 9.6 Individually georeferenced small- and large-scale sheets from Battista Agnese's portolan atlas (1538), overlay of their vectorised coastlines with the coastlines of their coherent composite, and the magnetic declination for the year 1500 according to the CALS3k.4 paleomagnetic model (Portolan atlas source: *University of Pennsylvania, Rare Book & Manuscript Library*, LJS 28). Basemap shapefile source: marineregions.org (Claus et al. 2017). CALS3k.4 data source: GEOMAGIA50.v3.2, accessed: March 18, 2023

shipborne use of the magnetic compass led to the creation of portolan charts based on contemporaneously acquired bearing observations (Fig. 9.6). Furthermore, Marelić's recent cartometric study of portolan atlases (Marelić 2024c, 2025a) shows that individual georeferencing of their sheets produces 'broken portolan chart images' because the anticlockwise tilts and map scales of their coastline renderings differ from those of their manually assembled coherent composites that were georeferenced in parallel (compare black and colour-coded vectorised tracings in Fig. 9.6) and that later-made atlases increasingly diverged from the contemporary magnetic declination values. These findings explicitly raise suspicion about the cartographers' in-depth knowledge of their own products and additionally decrease the likelihood of their genuine late medieval origin. Other significant pieces of cartometric evidence supporting the pre-medieval origin hypothesis include the fact that large portions of coastlines on the early, supposedly less-sophisticated *Pisane*, *Cortona*,

Avignon, and *Lucca* charts are very similar to each other and to Pietro Vesconte's charts, which are thought to be more mature and sophisticated (Marelić 2024a, 2025b); that portolan charts and atlases have not become more accurate over time (Marelić 2024b, 2025a); and that they exhibit twice as great planimetric accuracy after being split into their cartometrically determined sub-pieces that are georeferenced individually (see colour-coded vectorised tracings in Fig. 9.7). The last discovery was independently made by Marelić (2024a, b; 2025a, b) and Nicolai (2024), albeit through distinct methodological approaches. The most logical explanation for such phenomena is that late medieval cartographers were unable to properly assemble their copies of some earlier-made maps or charts (that they had somehow acquired in advance) and that they have stitched them together graphically with differences in scale and tilt.

Alongside cartometric analyses, various historical arguments challenge their medieval origin, including prolonged political upheavals in the region

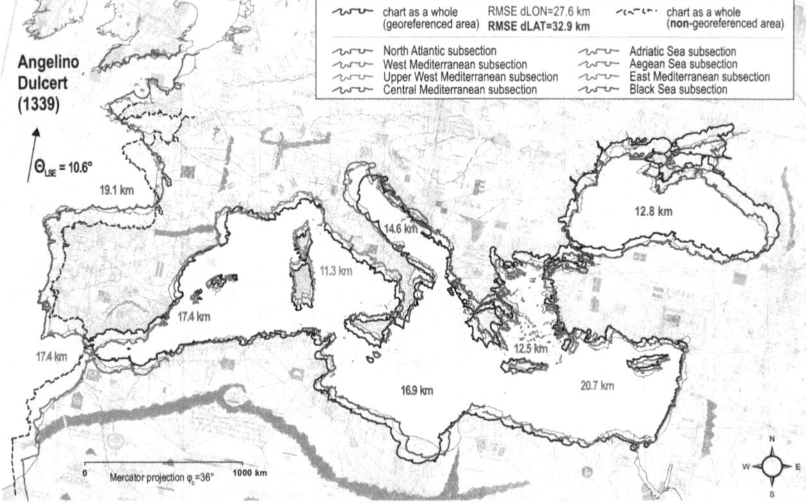

Fig. 9.7 Angelino Dulcert's portolan chart from 1339 georeferenced across the Mediterranean and Black Sea (black line) and its eight cartometrically determined subsections (colour-coded lines) with an average planimetric accuracy twice as high (Portolan chart source: *Bibliothèque nationale de France*, GE B-696 (RES)). Basemap shapefile source: marineregions.org (Claus et al. 2017)

(Abulafia 2011) that obstructed the practicality of extensive interstate campaigns for the hypothetical shipborne acquisition of substantial bearing and distance data, the pervasive lack of understanding of geodetic principles and map projection mechanics (as the earliest West European engagement with such concepts occurred through the rediscovery of Ptolemy's *Geography* during the early Renaissance), and the inadequacy of navigational instruments and sailboat manoeuvrability. Also, the earliest-known portolan charts were created decades before the invention and widespread adoption of the mariner's compass in the mid-fourteenth century, whereas the estimation of sailed distances emerged in the sixteenth and seventeenth centuries (Pujades 2007; Nicolai 2014). In contrast to the total lack of historical documentation validating the medieval origin hypothesis, three historical sources distinctly attribute their origins to classical antiquity, each created approximately a century apart. The earliest one is Qutb al-Din al-Shirazi's description (1282) of a map akin to a portolan chart, which he claims to be originally made by the sages of Greece and ancient geometers.[15] Two other historical records are the inscription on the anonymous *Rex Tholomeus* portolan chart (c. 1360),[16] and the passage from Benedetto Cotrugli's *De Navigatione Liber* (1464),[17] which unilaterally assert that portolan charts originated from Ptolemy's maps (Marelić 2024b). Marelić cartometrically established that the anticlockwise tilt of the Central Mediterranean on portolan charts corresponds to the tilt of the same area in the modern approximation of *Ptolemy's first projection* (equidistant conic projection) centred on the longitude of Gibraltar (see the upper part of Fig. 9.3). Also, the tilts of their subsections from west to east increase similarly to the tilts of the meridians on that map, in contrast to the magnetic declination of that period, which declined towards the east (Marelić 2024b). Moreover, Marelić's most recent cartometric study examines the geometry of al-Shirazi's data and some early-made portolan charts in conjunction with well-known historical sources and points at certain solid

[15] The additional information about Qutb al-Din al-Shirazi's records and their English transcript can be found in Savadi and Campbell (2023). For its detailed cartometric analysis that points at the classical antiquity origins of their spatial data, see Marelić (2025b).

[16] For more information regarding the *Rex Tholomeus* portolan chart, see Ruderman et al. (2023) and Marelić (2024b).

[17] Benedetto Cotrugli's *De Navigatione Liber* is the property of Yale University, Beinecke rare book and manuscript library, Call No.: MS 557 (https://collections.library.yale.edu/catalog/2005406). The passage appears on pages 60v (image ID: 1029839) and 61r (image ID: 1029840). Its transcript and English translation can be found in Marelić (2024b).

connections between the portolan charts and Claudius Ptolemy's geographical opus that were previously unimagined, including the first-ever objectively proposed reason for the length of the ***portolan mile*** (***miglio***) of approximately 1.25 km. Ptolemy's inaccurate dimensions of the Earth's circumference, which are 1.4 times smaller than the actual circumference, led to an incorrect division of the Mediterranean into about 60° of longitude instead of 42°. As a result, a one-degree arc along the parallel φ=36° mistakenly measures about 63 km, corresponding to 50 portolan miles. The lengths of the squares in the grids common to the *Pisane* and *Avignon* charts, as well as within al-Shirazi's matrix of 40×30 squares with a schematic appearance of the coastline remarkably similar to that on the *Pisane* and *Cortona* charts, amounting to 100 portolan miles, correspond to the length of a two-degree arc along the parallel φ=36° according to Ptolemy's inaccurate perception of the Earth's circumference (Marelić 2025b). These results, bolstered by historical evidence, cumulatively suggest that the typical geometry of portolan charts may be the result of a process in which late medieval cartographers created a graphical composite of copies of regional maps in cylindrical projection (made based on Eratosthenes's dimensions of the Earth) that they gradually tilted more and more from west to east in order to emulate the appearance of the Mediterranean on a map in conic projection rotated so that the meridian passing through the 'beginning of the Mediterranean' (its farthest west at the area of Gibraltar) is plotted vertically and provides a novel and more credible explanation of the geometry of portolan charts, favouring the hypothesis of classical antiquity origins. Therefore, they are not only critically significant for the advancement of the history of cartography as a discipline but also greatly enhance our understanding of human history. While their implications are predominantly debated within academic circles, their cultural relevance transcends this context and warrants greater dissemination to the general public.

References

Abulafia D (2011) The Great Sea: a human history of the Mediterranean. Oxford University Press, Oxford, New York

Berggren JL, Jones A (2000) Ptolemy's geography: an annotated translation of the theoretical chapters. Princeton University Press, Princeton

Billion P (2011) A newly discovered chart fragment from the Lucca Archives, Italy. Imago Mundi 63(1):1–33. https://doi.org/10.1080/03085694.2011.521326

Campbell T (1987) Portolan Charts from the Late Thirteenth Century to 1500. In: Harley JB, Woodward D (eds) The history of cartography, volume 1–cartography in Prehistoric, Ancient, and Medieval Europe and the Mediterranean. University of Chicago Press, Chicago, pp 371–463

Claus S, De Hauwere N, Vanhoorne B, Souza Dias F, Oset García P, Schepers L, Hernandez F, Mees J (Flanders Marine Institute) (2017). MarineRegions.org.

Clos-Arceduc A (1956) L'Énigme des Portulans: Étude sur la Projection et le Mode de Construction des Cartes à Rumbs du XIVe et XVe Siècle. Bulletin du Comité des Travaux Historiques et Scientifiques: Section de Géographie 69:215–231

Dilke OAW (1985) Greek and Roman Maps. Cornell University Press, Ithaca, New York

Duken AJ (1988) Reconstruction of the Portolan Chart of G. Carignano (c. 1310). Imago Mundi 40(1):86–95. https://doi.org/10.1080/03085698808592641

Edson E (2007) The world map 1300–1492. Johns Hopkins University Press

Fiorini M (1881) Le Projezioni delle carte geografiche. Bologna.

Gautier Dalché P (2007) The reception of Ptolemy's geography (End of the Fourteenth to Beginning of the Sixteenth Century). In: Woodward D (ed) The history of cartography, vol 3, Part 1 – Cartography in the European Renaissance. University of Chicago Press, Chicago, pp 285–364

GEOMAGIA50.v3.2 Archeomagnetic and Volcanic Query Form: https://geomagia.gfz-potsdam.de/geomagiav3/AAquery.php (accessed: March 18, 2023)

Jenny B (2006) MapAnalyst - a digital tool for the analysis of the planimetric accuracy of historical maps. e-Perimetron 1(3):239–245. ISSN 1790-3769

Jenny B (2010) New Features in MapAnalyst. e-Perimetron 5(3):176–180. ISSN 1790-3769

Jenny B, Hurni L (2011) Studying cartographic heritage: analysis and visualization of geometric distortions. Comput Graph 35:402–411. https://doi.org/10.1016/j.cag.2011.01.005

Jenny B, Weber A, Hurni L (2007) Visualizing the planimetric accuracy of historical maps with MapAnalyst. Cartographica 42(1):89–94. https://doi.org/10.3138/carto-v42-1-089

Kagan RL, Schmidt B (2007) Maps and the Early Modern state: official cartography. In: Woodward D (ed) The history of cartography, Part 1 – Cartography in the European Renaissance, vol 3. University of Chicago Press, Chicago, pp 661–679

Keuning J (1955) The history of geographical map projections until 1600. Imago Mundi 12(1):1–24. https://doi.org/10.1080/03085695508592085

Korte M, Constable C (2011) Improving geomagnetic field reconstructions for 0–3 ka. Phys Earth Planetary Interiors 188(3–4):247–259. https://doi.org/10.1016/j.pepi.2011.06.017

Livieratos E (2006) Graticule versus point positioning in Ptolemy cartographies. e-Perimetron 1(1):51–59. ISSN 1790-3769
Livieratos A, Boutoura C (2018) Carte Pisane and its coastline shape. e-Perimetron 13(3):161–181
Livieratos E, Tsorlini A, Boutoura C (2007) Coordinate analysis of Ptolemy's Geographia Europe Tabula X with respect to geographic graticule and point positioning in a Ptolemaic late 15th century map. e-Perimetron 2(2):80–91. ISSN 1790-3769
Livieratos E, Tsorlini A, Boutoura C, Manoledakis M (2008) Ptolemy's Geographia in digits. e-Perimetron 3(1):22–39. ISSN 1790-3769
Loomer SA (1987) A cartometric analysis of Portolan Charts: a search for methodology. Doctoral thesis, University of Wisconsin-Madison
Marelić T (2022) Conformal cylindrical properties of Adriatic Sea basin renderings on Portolan Charts. Cartographic J 59(2):83–101. https://doi.org/10.1080/00087041.2021.1956263
Marelić T (2023a) Geometric stalemate and de-evolution of Adriatic Sea representations on Early Modern Age Nautical Charts. Cartographic J 60(3):216–229. https://doi.org/10.1080/00087041.2023.2172533
Marelić T (2023b) Doubly perceived shape of the Adriatic Sea Basin on Early Modern Geographical Maps and Nautical Charts. Kartografija i geoinformacije 22(2):20–41. https://doi.org/10.32909/kg.22.39.2
Marelić T (2024a) Traces of the common origin of Carte Pisane, Cortona Chart and Pietro Vesconte's Charts. KN – J Cartography Geographic Inf 74(2):137–156. https://doi.org/10.1007/s42489-023-00154-6
Marelić T (2024b) Copying-Lineages of Portolan Chart Metrics and Implications of Their Pre-Medieval Origin. Int J Cartography (published online 16 July 2024): 1–26. https://doi.org/10.1080/23729333.2024.2349974
Marelić T (2024c) Mosaics within mosaics: the anatomy of Portolan Atlases. Abstracts Int Cartographic Assoc 7, 100: 1–2 https://doi.org/10.5194/ica-abs-7-100-2024
Marelić T (2025a) Mosaics within mosaics: the geometry of Portolan Atlases that Contradicts the Medieval Origin Hypothesis. KN – J Cartography Geographic Inf 75(1):57–78. https://doi.org/10.1007/s42489-024-00176-8
Marelić T (2025b) Are Portolan Charts and Portolan Mile Geometrically Rooted in Classical Antiquity? A Cartometric Analysis of al-Shirazi's "Greek Map" and the Pisane, Lucca, Avignon, and Cortona Charts. Cartographica 59(4):143–160. https://doi.org/10.3138/cart-2024-0029
Martínek J, Létal A (2023) Astronomically determined localities, the core part of Ptolemy's Geography. Journal of maps 19(1):1–10. https://doi.org/10.1080/17445647.2023.2195563
Marx C (2011) On the precision of Ptolemy's geographic coordinates in his Geographike Hyphegesis. Hist Geo- Space Sci 2(1):29–37. https://doi.org/10.5194/hgss-2-29-2011

Marx C (2012) Rectification of the ancient geographic coordinates in Ptolemy's Geographike Hyphegesis. Hist Geo- Space Sci 3(1):99–112. https://doi.org/10.5194/hgss-3-99-2012

Marx C (2016) The western coast of Africa in Ptolemy's Geography and the location of his prime meridian. Hist Geo- and Space Sci 7(1):27–52. https://doi.org/10.5194/hgss-7-27-2016

Mille J (2023) The Avignon Chart (c. 1300–c. 1310): an early attempt to represent the Northern Regions of Europe on a Nautical Chart. Imago Mundi 75(2):232–243. https://doi.org/10.1080/03085694.2023.2273677

Modenov PS, Parkhomenko AS (1965) Geometric transformations (English trans. Slater MBP). Academic Press, New York

Nicolai R (2014) A critical review of the hypothesis of a medieval origin for Portolan Charts. Doctoral thesis, Universiteit Utrecht

Nicolai R (2016) The enigma of the origin of portolan charts: a geodetic analysis of the hypothesis of a medieval origin. History of Science and Medicine Library, vol 52. Brill, Leiden

Nicolai R (2018) Analysing MapAnalyst and its application to portolan charts. e-Perimetron 13(3): 121-140. ISSN 1790-3769

Nicolai R (2024) The origin problem of nautical cartography: the importance of evidence and method. Int J Cartography 1-25. https://doi.org/10.1080/23729333.2024.2352822

Nordenskiöld AE (1897) Periplus: An essay on the early history of charts and sailing-directions (F. A. Bather, Trans.). P.A. Norstedt and Söner, Stockholm

Penzkofer M (2016) Geoinformatics. Books on Demand, Norderstedt

Pujades RJ (2007) Les Cartes Portolanes: la Representació Medievald'una Mar Solcada. Institut Cartogràfic de Catalunya, Barcelona

Ruderman B, Clausen A, Allport H (2023) The Rex Tholomeus portolan chart of 1360. Barry Lawrence Ruderman Antique Maps Inc., downloadable at: https://www.raremaps.com/gallery/detail/91710/the-rex-tholomeus-portolan-chart-anonymous (Accessed December 21, 2023)

Russo L (2004) The Forgotten Revolution: How Science Was Born in 300 BC and Why it Had to Be Reborn (trans. S. Levy). Springer Science and Business Media

Savadi F, Campbell T (2023) Qutb al-Dīn al-Shīrāzī's textual map of the Mediterranean Sea (1282) and its evident source in a portolan chart. Imago Mundi 75(2):199–231. https://doi.org/10.1080/03085694.2023.2275418

Skelton RA (1965) Looking at an early map. University of Kansas Libraries, Lawrence, Kansas

Snyder JP (1993) Flattening the Earth: two thousand years of map projections. The University of Chicago Press, Chicago and London

Snyder JP, Voxland PM (1989) An album of map projections, U.S. Geological Survey (USGS) Professional Paper 1453, USGS Federal Center, Denver

Stevenson EL (1991) Claudius Ptolemy: the geography. Dover Publications Inc, New York

Tibbets GR (1998) The beginnings of a cartographic tradition. In: Harley JB, Woodward D (eds) The history of cartography, Volume 2, Book 1 - Cartography in the Traditional Islamic and South Asian Societies. University of Chicago Press, Chicago, pp 90–107

Tobler W (1966) Medieval distortions: the projections of ancient maps. Annals of the Association of American Cartographers 56(2):351–360. https://doi.org/10.1111/j.1467-8306.1966.tb00562.x

Tupikova I (2014) Ptolemy's Circumference of the Earth, TOPOI – Towards a Historical Epistemology of Space, Max Planck Institute for the History of Sciences; Preprint: 464

Wagner H (1896, reprinted 1969) The Origin of the Medieval Italian Nautical Charts. Report of the Sixth International Geographical Congress London, Royal Geographical Society, 695–705, reprinted in Acta Cartographica 5: 476–485

Winter H (1948) the true position of Hermann Wagner in the controversy of the compass chart. Imago Mundi 5(1):21–26. https://doi.org/10.1080/03085694808591900

Winter H (1956) The Origin of the Sea Chart. Imago Mundi 13(1):39–44. https://doi.org/10.1080/03085695608592124

Woodward D (1987) Medieval Mappaemundi. In: Harley JB, Woodward D (eds) The History of Cartography, Volume 1–Cartography in Prehistoric, Ancient, and Medieval Europe and the Mediterranean. University of Chicago Press, Chicago, pp 286–370

Appendix

Nautical charts of the Adriatic Sea in the early modern period (with medieval prelude and modern postlude)

Author	*Title*	*Origin*	*Current location*
Manuscript nautical charts			
1. Pietro Vesconte	[Nautical chart of the Adriatic Sea]	Venice, 1318	Österreichische Nationalbibliothek, Vienna (hereinafter ÖNB), Cod. 594 (Cimel. 20), 10v-11r
2. Pietro Vesconte	[Nautical chart of the Adriatic Sea]	Venice, 1320	Vatican Library (Biblioteca Apostolica Vaticana), Rome; Cod. Pal. Lat.1362A
3. Pietro Vesconte	[Nautical chart of the Adriatic Sea]	Venice, 1321	Lyon Municipal Library (Bibliothèque Municipale de Lyon), Lyon; MS 175
4. Francesco de Cesanis	[Nautical chart of the Adriatic Sea]	Venice, 1421	Museo Civico Correr, Venezia (hereinafter MCC); Collezione Port. 13
5. Giacomo Giroldi (Ziroldi)	[Nautical chart of the Adriatic Sea]	Venice, 1426	Biblioteca Nazionale Marciana, Venice (hereinafter BNM); It. VI, 212 (=5694)

(*continued*)

J. Faričić, T. Marelić (eds.), *Early Modern Nautical Charts of the Adriatic Sea*, https://doi.org/10.1007/978-981-96-8046-7

(continued)

Author	Title	Origin	Current location
6. Grazioso Benincasa	[Nautical chart of the Adriatic Sea]	Venice, 1472	MCC, Gabinetto di Cartografia; Cl. XLIVa n. 5
7. Bartolommeo Zamberti (dalli Sonetti)	[Nautical chart of the Adriatic Sea], in: Isolario	Venice, 1485	National Maritime Museum, Greenwich, London (hereinafter NMM); P/21(2); MS 38-9920C
8. Zuan Soligo	[Nautical chart of the Adriatic Sea]	Venice, c. 1489	British Library, London; Egerton MS 73
9. Francesco de Cesanis	[Nautical chart of the Adriatic Sea]	Venice, c. 1489	British Library, London; Egerton MS 73
10. Vesconte Maggiolo	[Nautical chart of the Adriatic Sea]	Naples, 1511	John Carter Brown Library, Map Collection, Providence; 3-Size Codex Z 2
11. Pîrî Reis	[Nautical chart of the Adriatic Sea], in: Kitab-i bahriye	Gallipoli, 1526	The Walters Art Museum, Baltimore; W.658, fol. 208a
12. Battista Agnese	[Nautical chart of the Adriatic Sea]	Venice, 1538	University of Pennsylvania, Kislak Center for Special Collections, Rare Books and Manuscripts, Philadelphia; LJS 28
13. Alonso de Santa Cruz	[Nautical chart of the northern part of the Central Mediterranean]	Seville, c. 1545	National Library of Spain (Biblioteca Nacional de España), Madrid, MSS. MICRO/12638
14. Unknown author	[Nautical chart of the Adriatic Sea]	Dieppe, c. 1538–1546	Koninklijke Bibliotheek, Nationale bibliotheek, The Hague; 129 A 24
15. Unknown author	[Nautical chart of the Adriatic Sea]	Dieppe, 1547	The Huntington Library, Art Museum, and Botanical Gardens, Library Collections, Maps and Atlases, Portolans, San Marino, CA, USA (hereinafter HL); mssHM 29
16. Joan Martines	[Nautical chart of the Adriatic Sea]	Messina, c. 1550	NMM; P/25(5); MS 39 9926C

(continued)

(continued)

Author	*Title*	*Origin*	*Current location*
17. Diogo Homem	[Nautical chart of the Adriatic Sea]	Unknown place, 1559	BNF; CPL GE DD-2003 (RES), f. 6e
18. Hieronimo Masarachi	[Nautical chart of the Adriatic Sea]	Venice, c. 1560	Newberry Library, The Franco Novacco Map Collection, Chicago; Novacco 2R 1 (PrCt)
19. Diogo Homem	[Nautical chart of the Adriatic Sea]	Venice, 1570	Croatian State Archives, Cartographic Collection, Zagreb (hereinafter CSA); HR-HDA-902, D.XIV.6
20. Alī ibn Aḥmad Sharafī al-Ṣifāqsī	[Nautical chart of the Adriatic Sea]	Uknown place, c.1571	University of Oxford, Bodleian Library, Oxford; MS. Marsh 294, f. 6b
21. Diogo Homem	[Nautical chart of the Adriatic Sea]	Venice, 1574	CPL GE DD-2006 (RES), f. 4e
22. Antonio Millo	[Nautical chart of the Adriatic Sea]	Venice, 1583	The National Library of Poland, Warsaw; BN ZZK 0.2399
23. Vicko Dimitrije Voltius (Volčić)	Golfo di Venetia	Naples, 1593	National Library of Finland, Maps, The Nordenskiöld Map Collection, Helsinki; N-Kt-103c
24. Joan Oliva	[Nautical chart of the Adriatic Sea]	Marseille, 1613	British Library, London; Egerton MS 819
25. Alvise Gramolin	[Nautical chart of the Adriatic Sea]	Venice, 1624	MC; Cl. XLIVa n. 0044
26. Brasito Oliva	[Nautical chart of the Adriatic Sea]	Venice, 1633	BNM; It. IV, 126=5325
27. Pierre Collin	[Nautical chart of the Adriatic Sea]	Marseille, 1642	Bibliothèque Municipale de Lyon, Lyon; MS 177
28. Niccolò Guidalotti	[Nautical chart of the Adriatic Sea]	Venice, 1646	BNM; It. IV, 10=5062
29. Giovanni Battista Cavallini	Carta Particolare di G. di Venezia, in Teatro del Mondo Maritimo Conforme La Carta da Nauigare	Livorno, 1652	Institute and Museum of the History of Science, Florence; MED G.F 027

(*continued*)

(continued)

Author	*Title*	*Origin*	*Current location*
30. Placido Caloiro et Oliva,	[Nautical chart of the Adriatic Sea]	Venice, 1650	MCC; Cl. XLIVa n. 0011
31. Jean François Roussin	Carta du Golfo di Venetia	Venice, 1661	HL; mssHM 37
32. Marccheto Fassoi	[Nautical chart of the Adriatic Sea]	Venice, 1679	HL; mssHM 30
33. Jacob Robin	Pas Cart van Venedi	Unknown place, 1694	MCC; Cl. XLIVa n. 23
34. Filippo Francini	Principio del Golfo di Venezia, á Compasso Largo	Venice, 1699	ÖNB, Sammlung von Handschriften und alten Drucken; Cod. Ser. n. 12685, 7
35. Filippo Francini	[Nautical chart of the northern part of the Adriatic Sea]	Venice, 1699	ÖNB, Sammlung von Handschriften und alten Drucken; Cod. Ser. n. 12685, 6
36. Ferdinando Visconti	Carte generale ridotta del mare Adriatico	Milan, 1810	Private Collection Marco Asta, Bologna (hereinafter PC MA)
Printed nautical charts			
37. Pietro Coppo	Carta del Golfo Adrian	Venice, 1525	Maritime Museum Sergej Mašera, Piran
38. Giovanni Andrea Vavassore	La vera descrittione del Mare Adriatico: di L'arcipelago; e Mare di Soria...	Venice, 1541	NMM; G235:1/3
39. Paolo Forlani	Il Golfo di Venetia	Venice, 1568	BNF; GE CC-1380 (35RES)
40. William Barents	Tabula Hydrographica, In qua Italiae, orae maritimae; Item Venetiae, Istriae, Dalmatiae, Slavoniae, Graeciae..., in: Nieuwe beschryvinghe ende Caertboeck vande Midlandtsche Zee	Amsterdam, 1595	Stanford University Libraries, The Barry Lawrence Ruderman Map Collection, Stanford

(*continued*)

(continued)

Author	*Title*	*Origin*	*Current location*
41. Willem Janszoon Blaeu	Zeecaarte vande Golfe van Venetien, in: Tderde Deel vant Licht der Zee-vaer inhovdende de Beschryvinghe der Zee Kusten van de Middelantsche Zee	Amsterdam, 1621	National Library of Spain, Flemish and Dutch Atlases collections; GMG/527
42. Robert Dudly	Carta Particolare del mare Adriatico che comincia con il capo di Ancona è Finisce con L'Isola Lesina nello Isteso Mare	Florence, 1647	National Library of Finland, Helsinki; URN:NBN: fi-fe201002051338
43. Robert Dudly	Carta Particolare del mare Mediterraneo che comincia con Ciuita vecchia è Finisce con il Capo S. Maria in Calabria	Florence, 1647	National Library of Finland, Helsinki; URN:NBN: fi-fe201002051338
44. Robert Dudly	Carta Particolare del mare Mediterraneo che comincia con Budua in Dalmatia è Finisce con corfu nelo Stato Venetiano	Florence, 1647	National Library of Finland, Helsinki; URN:NBN: fi-fe201002051338
45. Pieter Goos	Zee-kaarte van de Golf van Venetien	Amsterdam, 1650	PC MA
46. Hendrick Doncker	Golfo van Venetien, in: De Lichtende Columne ofte Zee-Spiegel	Amsterdam, 1655	PC MA
47. Francesco Maria Levanto	Carta Maritima del Golfo di Venetia	Genoa, 1664	PC MA
48. Pierre du Val	[Nautical chart of the Adriatic Sea], in: La carte générale et les cartes particuliéres des costes de la mer Méditerranée	Paris, 1664	PC MA
49. John Seller	A Sea chart of the Gulph of Venice	London, 1677	CSA; HR-HDA-902, E.IV.35
50. Vincenzo Maria Coronelli	Golfo di Venezia	Venice, 1688	NUL; S-JZ-XVII-56

(*continued*)

(continued)

Author	*Title*	*Origin*	*Current location*
51. Ioannes van Keulen	Pas-Caart van de Weder zytsche Zee-kusten soo van Italia als Dalmatia en Griecken Inde Golff van Venetien	Amsterdam, 1700	CSA; HR-HDA-902, E. IV.30
52. Pierre Van der Aa	Golfe de Venise avec les Côtes maritimes, Bayes et Ports etc. De la Gréce, Dalmatie et Italie	Leiden, 1720	CSA; HR-HDA-902, E.IV.29
53. Joseph Roux	Carte de la mer Mediterranée, VII Feuille [Nautical chart of the Adriatic Sea]	Marseille, 1764	NUL; S-JZ-XVIII-116
54. Jacques-Nicolas Bellin	Carte hydrographique du Golphe de Venise	Paris, 1771	NUL; S-JZ-XVIII-145
55. Unknown author	[Carte turque du Golfe de Venise]	Unknown place, 18th c.	BNF; CPL GE DD-2987 (5853)
56. Lodovico Furlanetto	Nuova carta Marittima del Golfo di Venezia	Venice, 1784	State Archives in Zadar, Cartographic collection, Zadar; HR-DAZD-38, No. 3.1
57. Vincenzo di Lucio	[Nautical chart of the Adriatic Sea], 19 sheets	Venice, 1794.—1796.	PC MA
58. Giovanni Fileti	[Nautical chart of the Adriatic Sea], Foglio VI, in Carta Piana del mare Mediterraneo corretta, rettificata sotto i felicissimi auspici di Ferdinando IV, Re delle Due Sicilie per ordine del Signor Conte di Thurn e Valsassina	Naples, 1802	PC MA
59. Arcangelo Sartori	Nuova carta marittima del Golfo di Venezia	Ancona, 1802	PC MA
60. Giovanni Grubas	The Adriatic Sea or Gulf of Venice	London, 1803	British Library, Cartographic Items, Maps 20681.(6.)/Maps K.MAR.5

(*continued*)

(continued)

Author	Title	Origin	Current location
61. Direccion Hidrográfica	Carta esferica que comprehende Las Costas de Italia, Las del Mar Adriatico...	Madrid, 1804	PC MA
62. Giovanni Grubas	Nuova carta del Mare Adriatico ossia Golfo di Venezia	Trieste, 1809	Scientific Library, Zadar; 15188 D-20
63. Giovanni Grubas	Saggio di una Carta ridotta del Mare Adriatico ossia Golfo di Venezia	Trieste, 1816	PC MA
64. Georg Strudthoff	Chart of the Adriatic Sea or Gulf of Venice	Trieste, 1816	PC MA
65. Pierre Henri Gauttier	Carte Réduite du Golfe de Venise	Paris, 1820	PC MA
66. Military Geographic Institute	Carta di cabotaggio del Mare Adriatico (in twenty sheets)	Milan, 1822–1824	State Archives in Zadar, Cartographic collection, Zadar; HR-DAZD-383, No. 3.2
67. Direccion de Hidrografia	Carta Esferica que comprehende las Costas de Italia Las del Mar Adriatico	Madrid, 1824	Biblioteca Nacional de España, Madrid; MR/22 H. 100 N.2
68. Direccion de Hidrografia	Carta esférica de las Costas e islas de la mitad septentrional del Mar Adriático	Madrid, 1840	PC MA
69. Direccion de Hidrografia	Carta esférica de la mitad meridional del Mar Adriático	Madrid, 1842	PC MA

Index

A
Abruzzo, 214
Accuracy, planimetric, 4, 19, 225, 229, 232, 242
Adriatic, eastern, 204
Adriatic Sea, 1, 37, 39, 42, 51, 53, 57, 67, 69, 74, 80, 102, 106, 128, 131, 133, 134, 137, 141, 171, 179, 193, 198, 212, 217, 238
Affairs, maritime, 195
Africa, 150, 155
Agnese, Battista, 14, 106
Ahmed the Tangier, 199
Albania, 44, 57, 216
Ali Macar Reis, 207, 208, 216
Allegory, 76–78
Alps, 102, 108, 150
Amphitrite, 160
Amsterdam, 162
Analogy (method), 20
Analysis
  cartometric, 20, 26, 223, 226–229, 232
  comparative, 25
  qualitative, 22
  quantitative, 22
Anatolia, 198
Anchorage, 85, 86
Ancona, 37, 45, 46, 48, 50, 53, 58, 205, 212
  March of, 214
Angel, 159, 162, 164
Antipaxos, 216
Antiquity, classical, 4, 24
Apennines, 150
Apollonius, 148
Approach, interdisciplinary, 124
Arcadia, 134, 135
Argonauts, 148
Art, fine, 149
Ascension, Feast of the, 160
Asia, 150, 156
Astrolabe, 162
Astronomy, 162
Athos, 214
Atlas, 211
  portolan, 241

B
Bar, 214, 216
Barents, William, 85, 87, 89

J. Faričić, T. Marelić (eds.), *Early Modern Nautical Charts of the Adriatic Sea*, https://doi.org/10.1007/978-981-96-8046-7

Barentsz, Willem, 12
Baroque, 149, 159
Bay of Pag, 175
Beautemps-Beaupré, Charles François, 18
Belgrade, 44, 57
Bellin, Jacques Nicolas, 14
Benincasa, Grazioso, 8, 104
Bertelli, Andrea, 183
Biograd, 178
Bjelovar, 214
Black Sea, 148, 199
Blaeu, Joan, 197, 214
Blaeu, Willem Janszoon, 12, 89
Bologna, 159
Bonifacio, Natale, 198
Bonomini, Bartolomeo, 49, 51, 52
Bordone, Benedetto, 107–108, 170, 173, 184
Bosnia, 52, 54, 208, 214
Brač, 176, 205, 213
Brusnik, 216
Budva, 216
Buondelmonti, Cristoforo, 107, 169

C
Camocio, Giovanni Francesco, 108, 171, 178, 179, 204
Carnia, 203
Carte da navigare, 103, 104
Carte da terra, 103, 108, 109
Cartographic metrics, 109
Cartographic school, Mallorcan, 199
Cartography, 147, 148, 151, 162, 195, 197
  European, 193, 196, 212
  nautical, 69
  Ottoman, 195, 196, 200, 208, 215, 217
Cartouche, 162
Çelebi, Evliya, 195
Çelebi, Kâtib, 196, 197, 212, 214, 215, 217, 218
Chart
  manuscript, 9, 79
  nautical, 4, 68, 69, 74, 78, 82, 86, 99, 103, 104, 106, 123, 127, 130, 135, 142, 179, 181, 194, 198
  portolan, 4, 8, 14, 26, 79, 87, 172, 174, 199, 224, 225, 233, 240, 243
  printed, 79, 80
Chartmakers, 42, 52, 57
Christianity, 150, 152, 155, 164
Chronometer, 7
Church, 160, 161
  Catholic, 160
Coast, Adriatic, 208
Coastline, Adriatic, 198
Collin, Pierre, 160
Colonna, Marcantonio di Pagliano, 50, 51
Communication, 67, 71–72
  cartographic, 9, 20, 23
Compass, magnetic, 241
Compass rose, 6, 152, 154, 155, 163
Compilation (uncritical), 12
Computer, 8, 26
Condina, Sebastiano, 149
Constantinople, 194
Control, territorial, 124, 127
Coordinate Reference System (CRS), 231, 234, 236
Coordinates, 226, 227, 236, 238
  spherical, 4
Corfu, 102, 108, 216
Coronelli, Vincenzo Maria, 2, 15, 108, 128, 130, 142, 148, 152, 159, 171, 184, 204
Cotrugli, Benedetto, 7, 243
Council of Ten, 111, 113, 209
Counties (Comitatus, županije), 53

Creature, exotic, 133, 143
Cres, 148, 173, 176
Cresques, Abraham, 199
Croatia, 39, 41, 44, 52, 54, 214
Cross, 152, 154, 155

D
Đakovo, 214
Dalmatia, 38, 39, 43, 45, 46, 52–54, 57, 60, 102, 108, 130, 159, 205, 214
Data
  identical, 227
  reference, 227
De Aguiar, Jorge, 152
De Cesanis, Francesco, 8
Declination, magnetic, 7, 15, 240
Decomposition (method), 20, 25
Deconstruction (method), 20, 25, 69, 73, 74
Decoration, 149
De Lucio, Vincenzo, 18, 77, 78, 85, 89
De-projecting, 228, 232, 235
Depth, 85
De Santa Cruz, Alonso, 171
Description (method), 22
Al-Din al-Shirazi, Qutb, 243
Dinarides, 150
Discourse, geopolitical, 81
Dogado, 108
Doge, 37, 39, 53, 55, 60, 100, 101
Dominio, 102, 103, 108–111, 113
Dubrovnik, 44, 53, 157, 204, 205, 208, 213, 214, 216
Dudley, Robert, 12
Dulcert, Angelino, 199
Đurđevac, 214
Durrës, 208, 214, 216

E
Early modern period, 124, 125, 135
Eastern Adriatic, 172, 176
East, Middle, 155
Ebubekir bin Behram ed-Dımeşkî Efendi, 197, 214
Element
  decorative, 124, 125, 127, 135
  imagological, 125
Europe, 150
Exchange, cultural, 210

F
Fassoi, Marcheto, 14
Ferretti family, 48, 49, 51
Fine, Oronce, 210
Flag, 75
Fleur-de-lis, 152
Florence, 169
Forest, 102, 112, 113, 115
Forlani, Paolo, 106, 107, 110
Franco, Giacomo, 171, 184
Freducci family, 41, 45, 46, 48
Furlanetto, Ludovico, 16–18, 104

G
Gastaldi, Giacomo, 101, 210
Generalisation, cartographic, 170, 173
Genoa, 151
Geographic Information System (GIS), 8, 27, 227, 236
Geography, 194, 197
  Islamic, 199
Geometry
  ellipsoidal, 231, 234
  Euclidean, 231, 234
  spherical, 4, 231, 234
Geopolitics, 100, 103, 105, 108, 109
Georeferencing, 27, 232, 233

Giroldi, Giacomo, 8
Golfo di Venezia, 68, 80
Gouvia, 216
Gramolin, Alvise, 157
Graticule, 6, 13
Greuter, Mattheus, 159
Griselini, Francesco, 101
Guild of Rumi Painters, 208
Gulf of Adriatic, 202
Gulf of Venice, 214, 216

H
Habsburg Empire, 204
Al-Hac Ebu'l-Hasan, 199
Hajji Ahmed of Tunisia, 209
Heraldry, 75–76
Herceg Novi, 216
Herzegovina, 216
Hezârfen Hüseyin Efendi, 197
Hierarchy (of the rhetoric), 127, 135–141
Hipparchus, 200
Holy Trinity, 153
Homem, Diogo, 14
Hüseyin Efendi, 197
Hvar, 176, 181, 213, 216

I
Ibrahim al-Mursi, 199
Iconography, 125, 131–135
Identity
  Christian, 138
  cultural, 131
  religious, 130
Al-Idrisi, Muhammad, 154
Ihlasi, Sheikh Mehmed, 196
Imagery, 129–131, 138, 142
Imagology, 20, 74–78
Information
  cartographic, 71, 72
  geographical, 195
Ionian Sea, 2, 157
Islam, 150, 154, 155
Island, 82, 89
Isolario, 8, 103, 106, 108, 151, 169, 171
Ist, 89
Istanbul, 154, 218
Istra, 38, 40, 49, 57
Istria, 108, 203, 214
Italy, 38, 43, 46, 52, 58, 159, 210, 212, 214, 216

J
Jablanac, 216
Jerusalem, 154
Jesus, 149, 153, 154

K
Kanjiža, 214
Al-Kindī, Abu Yusuf Ya'qub ibn Ishaq, 201
Kingdom of Naples, 2, 156
Kingdom of Sicily, 156
Klenau, Giovanni von, 39, 40
Klis, 188
Koločep, 205
Konavle, 214
Koper, 208
Korčula, 181, 205, 213, 216
Kornat, 213
Kotor, 208, 214, 216
Krk, 173, 176
Kvarner, 184
Kvarner Gulf, 173
Kvarner Islands, 148

L
Landscape
  cultural, 127
  pastoral, 133
Lastovo, 181
Latitude, 4, 10, 225, 236
League, 88
Least Squares Estimation (LSE), 228
Leiden, 160
Leonardi, Antonio, 101
Lepanto, 39, 49, 50, 52, 55
Leschiutta, Francesco, 115
Lezhë, 214, 216
Licini, family, 39, 54, 56, 60
Ljubač, 216
Lokrum, 81
Lonati, Abbate, 106, 107
Longitude, 4, 7, 11, 225, 236, 237
Loreto, 45, 50, 53

M
Mainland, 101, 105, 106, 111, 112, 118
Map, 68, 70, 71, 73, 194
  chorographic, 103, 105, 106
  geographical, 11
  old, 149
  scale, 82, 86, 87
MapAnalyst (software), 27, 232
Mapping distortions, 231, 235
Mapping errors, 229
Mare Adriaticum, 80
Marieni, Giacomo, 19, 204
Marinos of Tyre, 201
Marseille, 160
Martines, Joan, 154
Mediterranean Sea, 1, 102, 150, 169, 198, 199, 211, 225
Medium, 124
  of communication, 70, 73–74
Mercator, Gerardus, 196
Merian, Matthäus, 198
Meridian, prime, 237
Message, 68, 69, 71, 75, 80, 91
Messina, 154
Milan, 155, 212
Mile, 87, 88
Military Order of St Stephen, 45, 48, 51, 57
Millo, Antonio, 75, 155, 171, 174
Mljet, 213
Model (of communication), 71, 72, 91
Moon, crescent, 152, 154, 155
Motif, religious, 148, 149, 151, 164
Mythology, Greek, 148

N
Naming, 74, 80–82, 89
Naples, 149, 151
Napoleon, 37, 41, 58
Navigation, 69, 82, 83, 85, 149, 162, 195
Neptune, 159
Neretva, 213
New World, 198, 210
Nin, 208, 216
Noise, communication, 89–91
North Dalmatia, 173

O
Olib, 175
Ornament, exotic, 137
Osor, 148
Other, 74, 131–135, 142, 143
Ottoman Empire, 2, 39, 43, 49, 52, 54, 128, 132, 133, 135, 138, 154, 155, 164, 193, 194, 196, 204, 210, 214, 217

P

Pag, 173, 174
Pagano, Matteo, 110
Pakrac, 214
Palagruža, 216
Pannonia, 203
Papal States, 156, 164
Parallel, true-to-scale, 235
Pareto, Bartolomeo, 42
Pašman, 179
Paxos, 216
Pelješac Peninsula, 182, 216
Peninsula, Apennine, 161
Period, early modern, 176
Periplous, 4
Pinargenti, Simone, 171, 204
Piombo (lead ore) in Dalmatia, 54, 56, 60
Pîrî Reis, Muhiddin, 13, 153, 171, 179, 195, 196, 204, 205, 212, 216, 218
Planudes, 202
Poglizzani (people), 53, 54, 56, 60
Polače, 214
Polionymy, 80
Pomena, 214
Popes, 37, 43–45, 49–51, 57
Porcachi, Tomaso, 108, 171
Poreč, 203
Portolan, 4, 169
Portolano, 151, 172
Portorož, 208
Poseidon, 159
Potential, communicative, 74, 81, 86, 88
Premantura, 216
Premuda, 89
Printing
  copperplate, 9
  woodcut, 9
Projection
  map, 4, 225, 231, 234
  Mercator, 6, 7, 13
Proversa, 213, 216
Ptolemy, Claudius, 10, 174, 195, 200, 225, 237, 238, 244
Pula, 203, 204

Q

Quarner, 102

R

Rab, 173, 176
Ragusa, 216
Ramusio, Giovan Battista, 101
Religion, 147, 148
  Greek and Roman, 159, 160, 164
Renaissance, 10, 159, 174, 177, 226, 243
Representation, 68, 71, 73, 74, 78
  anachronistic, 16
  cartographic, 16
Republic of Ragusa, 2, 156
Republic of Venice, 1, 51, 53, 55, 57, 68, 78, 81, 91, 100, 102, 104, 127, 136, 138, 141, 142, 157, 164, 171, 204
Residuals, 228, 235
Rhumb line, 7, 231
Rialto Bridge, 205
Rijeka, 216
Rock, 83
  awash, 83
Romagna, Papal State, 37, 38
Rome, 160
Root Mean Square Error (RMSE), 228, 232
Rosaccio, Giuseppe, 108, 171, 174, 180, 181, 184, 204
Rota Kolunić, Martin, 178, 198
Rouissin, Jean François, 157
Route, maritime, 204
Roux, Joseph, 14

S
Sabbadino, Cristoforo, 113
St. Blaise, 156
St. Cyriacus, 45
St. John, 157
St. Luke, 157
St. Mark, 155, 157, 159
St. Mary, 149
St. Matthew, 157
St. Nicholas, 149
Saint, patron, 125, 141
Sala dello Scudo, 100
Sarande, 216
Sava River, 203
Sazan, 214
Scale bar, linear, 6
Šćedro, 205
Scolari, Stefano, 109, 110
Sea depth marks, 14, 18
Seafaring, 3
Sea monster, 136, 141
Seller, John, 89
Senj, 216
Serbia, 45, 53
Serenissima, 102, 103, 105, 108, 109, 111, 113, 116
Shoal, 83
Šibenik, 39, 53, 54, 56, 216
Sign, cartographic, 83, 85
Skradin, 216
Slavonia, 214
Slavoni (Schiavoni, Sclavoni), 39, 44, 52, 55, 58
Smokvica, 208, 213, 216
Sobra, 213
Šolta, 205
Sorte, Cristoforo, 101, 113, 114
Space, 147, 148, 151
  profane, 148
  sacred, 148
Spinelli, Giacomo, 116, 117
Split, 180, 208, 213, 214, 216
Staff, Jacob's, 162
Stato da Mar, 39, 43, 54, 55, 57
Stato da Terra, 39, 55
Strait of Osor, 204
Strait of Otranto, 102, 106
Suhopolje, 214
Survey, 171, 177
  geodetic, 8
  hydrographic, 8, 18, 22, 85–87
Sušac, 205
Sv. Andrija, 205, 208, 216
Symbol
  generic, 136
  religious, 129, 130, 155, 160, 163
Symbolism, 100, 101, 109, 118, 125, 130, 149, 155

T
Technique, engraving printing, 170
Tentivo, Gaspare, 104, 105
Territory, Hungarian, 214
Theory, communication, 69, 71
Thevet, André, 171, 174, 182
Tiara, papal, 160, 164
Toponym, 6, 78–82, 89, 183, 184
Toponymy, Turkish, 211
Toreta, 213
Transformation
  affine, 233
  geometric, 228, 229, 232
  Helmert, 233
  projective, 233
Travelogue, 195
Treaty of Karlowitz, 194
Trieste, 203
Triton, 160
Trogir, 180, 181, 183, 216
Trsteno, 205

U
Ugljan, 179
Ulcinj, 216
Unit of distance, 87, 88
Urbino-Pesaro, Papal Duchy, 49, 51, 53
Uskoks, 176

V
Van der Aa, Pieter, 160
Van Keulen, Ioannes, 162
Vavassore, Giovanni Andrea, 106
Veduta, 125, 129, 133, 138, 140
Velebit Channel, 176
Velebit Mountain, 179
Vellum, 4
Venice, 37–39, 41, 50, 52, 53, 55, 58, 151, 155, 157, 159, 160, 169, 171, 203, 205, 208, 212
Vesconte, Pietro, 8, 83, 89, 104, 198, 227, 230, 233
Vir, 175, 179
Virovitica, 214
Vis, 188, 216
Visconti, Ferdinando, 18
Vlora, 208, 212, 214, 216
Vlora River, 208
Volčić, Vicko Dimitrije, 53
Voltius, Vincus Demetrius, 14

W
Wagner, Hermann, 227, 240
Water, 102, 112, 113
Western Europe, 174
Wind rose, 6, 199
Winged lion, 75, 77, 78, 130, 131, 138, 139, 141, 155, 157, 159, 160, 164

Z
Zadar, 38, 41, 46, 53, 60, 178, 208, 216
Zamberti da li Sonetti, Bartolommeo, 8, 107
Zorzi, Giovanni Domenico, 101

The manufacturer's authorised representative in the EU is Springer Nature Customer Service Centre GmbH, Europaplatz 3, 69115 Heidelberg, Germany. If you have any concerns regarding our products, please contact ProductSafety@springernature.com

Printed and bound by CPI Group (UK) Ltd, Croydon, CR0 4YY
01/10/2026
02225912-0007